DOING MATH WITH PYTHON

DOING MATH WITH PYTHON

Use Programming to Explore Algebra, Statistics, Calculus, and More!

by Amit Saha

no starch press

San Francisco

Printed in USA

First printing

19 18 17 16 15   1 2 3 4 5 6 7 8 9

ISBN-10: 1-59327-640-0
ISBN-13: 978-1-59327-640-9

Text stock is SFI Certified

Publisher: William Pollock
Production Editor: Riley Hoffman
Cover Illustration: Josh Ellingson
Interior Design: Octopod Studios
Developmental Editors: Seph Kramer and Tyler Ortman
Technical Reviewer: Jeremy Kun
Copyeditor: Julianne Jigour
Compositor: Riley Hoffman
Proofreader: Paula L. Fleming

For information on distribution, translations, or bulk sales, please contact No Starch Press, Inc. directly:

No Starch Press, Inc.
245 8th Street, San Francisco, CA 94103
phone: 415.863.9900; info@nostarch.com
www.nostarch.com

Library of Congress Cataloging-in-Publication Data

Saha, Amit, author.
  Doing math with Python : use programming to explore algebra, statistics, calculus, and more! / by Amit Saha.
        pages cm
  Summary: "Uses the Python programming language as a tool to explore high school-level mathematics like statistics, geometry, probability, and calculus by writing programs to find derivatives, solve equations graphically, manipulate algebraic expressions, and examine projectile motion. Covers programming concepts including using functions, handling user input, and reading and manipulating data"-- Provided by publisher.
  Includes index.
  ISBN 978-1-59327-640-9 -- ISBN 1-59327-640-0
  1. Mathematics--Study and teaching--Data processing. 2. Python (Computer program language) 3. Computer programming. I. Title.
  QA20.C65S24 2015
  510.285'5133--dc23
                                        2015009186

To Protyusha, for never giving up on me

BRIEF CONTENTS

CONTENTS IN DETAIL

3
DESCRIBING DATA WITH STATISTICS   61

4
ALGEBRA AND SYMBOLIC MATH WITH SYMPY   93

ACKNOWLEDGMENTS

I would like to thank everyone at No Starch Press for making this book possible. From the first emails discussing the book idea with Bill Pollock and Tyler Ortman, through the rest of the process, everyone there has been an absolute pleasure to work with. Seph Kramer was amazing with his technical insights and suggestions and Riley Hoffman was meticulous in checking and re-checking that everything was correct. It is only fair to say that without these two fine people, this book wouldn't have been close to what it is. Thanks to Jeremy Kun and Otis Chodosh for their insights and making sure all the math made sense. I would also like to thank the copy-editor, Julianne Jigour, for her thoroughness.

SymPy forms a core part of many chapters in this book and I would like to thank everyone on the SymPy mailing list for answering my queries patiently and reviewing my patches with promptness. I would also like to thank the matplotlib community for answering and clearing up my doubts.

I would like to thank David Ash for lending me his Macbook, which helped me when writing the software installation instructions.

I also must thank every writer and thinker who inspired me to write, from humble web pages to my favorite books.

INTRODUCTION

This book's goal is to bring together three topics near to my heart—programming, math, and science. What does that mean exactly? Within these pages, we'll programmatically explore high school–level topics, like manipulating units of measurement; examining projectile motion; calculating mean, median, and mode; determining linear correlation; solving algebraic equations; describing the motion of a simple pendulum; simulating dice games; creating geometric shapes; and finding the limits, derivatives, and integrals of functions. These are familiar topics for many, but instead of using pen and paper, we'll use our computer to explore them.

We'll write programs that will take numbers and formulas as input, do the tedious calculations needed, and then spit out the solution or draw a graph. Some of these programs are powerful calculators for solving math problems. They find the solutions to equations, calculate the correlation between sets of data, and determine the maximum value of a function, among other tasks. In other programs, we'll simulate real-life events, such as projectile motion, a coin toss, or a die roll. Using programs to simulate such events gives us an easy way to analyze and learn more about them.

You'll also find topics that would be extremely difficult to explore without programs. For example, drawing fractals by hand is tedious at best and close to impossible at worst. With a program, all we need to do is run a for loop with the relevant operation in the body of the loop.

I think you'll find that this new context for "doing math" makes learning both programming and math more exciting, fun, and rewarding.

Who Should Read This Book

If you yourself are learning programming, you'll appreciate how this book demonstrates ways to solve problems with computers. Likewise, if you teach such learners, I hope you find this book useful to demonstrate the application of programming skills beyond the sometimes abstract world of computer science.

This book assumes the reader knows the absolute basics of Python programming using Python 3—specifically, what a function is, function arguments, the concept of a Python class and class objects, and loops. Appendix B covers some of the other Python topics that are used by the programs, but this book doesn't assume knowledge of these additional topics. If you find yourself needing more background, I recommend reading *Python for Kids* by Jason Briggs (No Starch Press, 2013).

What's in This Book?

This book consists of seven chapters and two appendices. Each chapter ends with challenges for the reader. I recommend giving these a try, as there's much to learn from trying to write your own original programs. Some of these challenges will ask you to explore new topics, which is a great way to enhance your learning.

- **Chapter 1**, **Working with Numbers**, starts off with basic mathematical operations and gradually moves on to topics requiring a higher level of math know-how.
- **Chapter 2**, **Visualizing Data with Graphs**, discusses creating graphs from data sets using the matplotlib library.
- **Chapter 3**, **Describing Data with Statistics**, continues the theme of processing data sets, covering basic statistical concepts—mean, median, mode, and the linear correlation of variables in a data set. You'll also learn to handle data from CSV files, a popular file format for distributing data sets.

- **Chapter 4, Algebra and Symbolic Math with SymPy**, introduces symbolic math using the SymPy library. It begins with the basics of representing and manipulating algebraic expressions before introducing more complicated matters, such as solving equations.

- **Chapter 5, Playing with Sets and Probability**, discusses the representation of mathematical sets and moves on to basic discrete probability. You'll also learn to simulate uniform and nonuniform random events.

- **Chapter 6, Drawing Geometric Shapes and Fractals**, discusses using matplotlib to draw geometric shapes and fractals and create animated figures.

- **Chapter 7, Solving Calculus Problems**, discusses some of the mathematical functions available in the Python standard library and SymPy and then introduces you to solving calculus problems.

- **Appendix A, Software Installation**, covers installation of Python 3, matplotlib, and SymPy on Microsoft Windows, Linux, and Mac OS X.

- **Appendix B, Overview of Python Topics**, discusses several Python topics that may be helpful for beginners.

Scripts, Solutions, and Hints

This book's companion site is *http://www.nostarch.com/doingmathwithpython/*. Here, you can download all the programs in this book as well as hints and solutions for the challenges. You'll also find links to additional math, science, and Python resources I find useful as well as any corrections or updates to the book itself.

Software is always changing; a new release of Python, SymPy, or matplotlib may cause a certain functionality demonstrated in this book to behave differently. You'll find any of these changes noted on the website.

I hope this book makes your journey into computer programming more fun and immediately relevant. Let's do some math!

1

WORKING WITH NUMBERS

Let's take our first steps toward using Python to explore the world of math and science. We'll keep it simple now so you can get a handle on using Python itself. We'll start by performing basic mathematical operations, and then we'll write simple programs for manipulating and understanding numbers. Let's get started!

Basic Mathematical Operations

The *Python interactive shell* is going to be our friend in this book. Start the Python 3 IDLE shell and say "hello" (see Figure 1-1) by typing print('Hello IDLE') and then pressing ENTER. (For instructions on how to install Python and start IDLE, see Appendix A.) IDLE obeys your command and prints the words back to the screen. Congratulations—you just wrote a program!

When you see the >>> prompt again, IDLE is ready for more instructions.

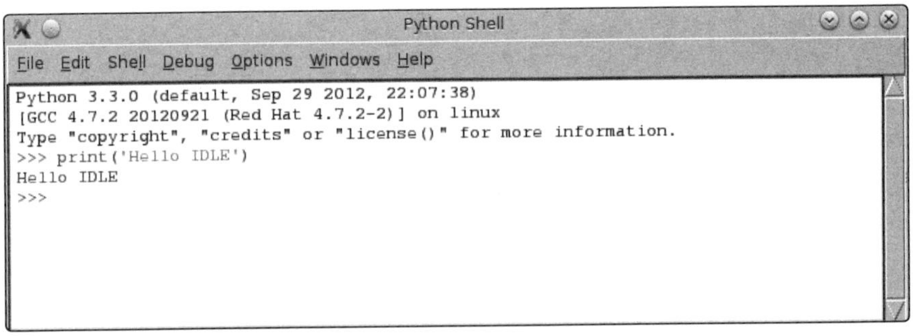

Figure 1-1: Python 3 IDLE shell

Python can act like a glorified calculator, doing simple computations. Just type an expression and Python will evaluate it. After you press ENTER, the result appears immediately.

Give it a try. You can add and subtract numbers using the addition (+) and subtraction (-) operators. For example:

```
>>> 1 + 2
3
>>> 1 + 3.5
4.5
>>> -1 + 2.5
1.5
>>> 100 - 45
55
>>> -1.1 + 5
3.9
```

To multiply, use the multiplication (*) operator:

```
>>> 3 * 2
6
>>> 3.5 * 1.5
5.25
```

To divide, use the division (/) operator:

```
>>> 3 / 2
1.5
>>> 4 / 2
2.0
```

As you can see, when you ask Python to perform a division operation, it returns the fractional part of the number as well. If you want the result in the form of an integer, with any decimal values removed, you should use the floor division (//) operator:

```
>>> 3 // 2
1
```

The floor division operator divides the first number by the second number and then rounds down the result to the next lowest integer. This becomes interesting when one of the numbers is negative. For example:

```
>>> -3 // 2
-2
```

The final result is the integer lower than the result of the division operation (-3/2 = -1.5, so the final result is -2).

On the other hand, if you want just the remainder, you should use the modulo (%) operator:

```
>>> 9 % 2
1
```

You can calculate the power of numbers using the exponential (**) operator. The examples below illustrate this:

```
>>> 2 ** 2
4
>>> 2 ** 10
1024
>>> 1 ** 10
1
```

We can also use the exponential symbol to calculate powers less than 1. For example, the square root of a number n can be expressed as $n^{1/2}$ and the cube root as $n^{1/3}$:

```
>>> 8 ** (1/3)
2.0
```

As this example shows, you can use parentheses to combine mathematical operations into more complicated expressions. Python will evaluate the expression following the standard *PEMDAS* rule for the order of calculations—parentheses, exponents, multiplication, division, addition, and subtraction. Consider the following two expressions—one without parentheses and one with:

```
>>> 5 + 5 * 5
30
>>> (5 + 5) * 5
50
```

In the first example, Python calculates the multiplication first: 5 times 5 is 25; 25 plus 5 is 30. In the second example, the expression within the parentheses is evaluated first, just as we'd expect: 5 plus 5 is 10; 10 times 5 is 50.

These are the absolute basics of manipulating numbers in Python. Let's now learn how we can assign names to numbers.

Labels: Attaching Names to Numbers

As we start designing more complex Python programs, we'll assign names to numbers—at times for convenience, but mostly out of necessity. Here's a simple example:

```
❶ >>> a = 3
   >>> a + 1
   4
❷ >>> a = 5
   >>> a + 1
   6
```

At ❶, we assign the name a to the number 3. When we ask Python to evaluate the result of the expression a + 1, it sees that the number that a refers to is 3, and then it adds 1 and displays the output (4). At ❷, we change the value of a to 5, and this is reflected in the second addition operation. Using the name a is convenient because you can simply change the number that a points to and Python uses this new value when a is referred to anywhere after that.

This kind of name is called a *label*. You may have been introduced to the term *variable* to describe the same idea elsewhere. However, considering that *variable* is also a mathematical term (used to refer to something like x in the equation $x + 2 = 3$), in this book I use the term *variable* only in the context of mathematical equations and expressions.

Different Kinds of Numbers

You may have noticed that I've used two kinds of numbers to demonstrate the mathematical operations—numbers without a decimal point, which you already know as *integers*, and numbers with a decimal point, which programmers call *floating point numbers*. We humans have no trouble recognizing and working with numbers whether they're written as integers, floating point decimals, fractions, or roman numerals. But in some of the programs that we write in this book, it will only make sense to perform a task on a particular type of number, so we'll often have to write a bit of code to have the programs check whether the numbers we input are of the right type.

Python considers integers and floating point numbers to be different *types*. If you use the function type(), Python will tell you what kind of number you've just input. For example:

```
>>> type(3)
<class 'int'>

>>> type(3.5)
<class 'float'>

>>> type(3.0)
<class 'float'>
```

Here, you can see that Python classifies the number 3 as an integer (type 'int') but classifies 3.0 as a floating point number (type 'float'). We all know that 3 and 3.0 are mathematically equivalent, but in many situations, Python will treat these two numbers differently because they are two different types.

Some of the programs we write in this chapter will work properly only with an integer as an input. As we just saw, Python won't recognize a number like 1.0 or 4.0 as an integer, so if we want to accept numbers like that as valid input in these programs, we'll have to convert them from floating point numbers to integers. Luckily, there's a function built in to Python that does just that:

```
>>> int(3.8)
3
>>> int(3.0)
3
```

The function int() takes the input floating point number, gets rid of anything that comes after the decimal point, and returns the resulting integer. The float() function works similarly to perform the reverse conversion:

```
>>> float(3)
3.0
```

float() takes the integer that was input and adds a decimal point to turn it into a floating point number.

Working with Fractions

Python can also handle fractions, but to do that, we'll need to use Python's fractions module. You can think of a *module* as a program written by someone else that you can use in your own programs. A module can include classes, functions, and even label definitions. It can be part of Python's standard library or distributed from a third-party location. In the latter case, you would have to install the module before you could use it.

The fractions module is part of the standard library, meaning that it's already installed. It defines a class Fraction, which is what we'll use to enter fractions into our programs. Before we can use it, we'll need to *import* it, which is a way of telling Python that we want to use the class from this module. Let's see a quick example—we'll create a new label, f, which refers to the fraction 3/4:

```
❶ >>> from fractions import Fraction
❷ >>> f = Fraction(3, 4)
❸ >>> f
Fraction(3, 4)
```

We first import the Fraction class from the fractions module ❶. Next, we create an object of this class by passing the numerator and denominator as parameters ❷. This creates a `Fraction` object for the fraction 3/4. When we print the object ❸, Python displays the fraction in the form Fraction(*numerator, denominator*).

The basic mathematical operations, including the comparison operations, are all valid for fractions. You can also combine a fraction, an integer, and a floating point number in a single expression:

```
>>> Fraction(3, 4) + 1 + 1.5
3.25
```

When you have a floating point number in an expression, the result of the expression is returned as a floating point number.

On the other hand, when you have only a fraction and an integer in the expression, the result is a fraction, even if the result has a denominator of 1.

```
>>> Fraction(3, 4) + 1 + Fraction(1/4)
Fraction(2, 1)
```

Now you know the basics of working with fractions in Python. Let's move on to a different kind of number.

Complex Numbers

The numbers we've seen so far are the so-called *real numbers*. Python also supports *complex numbers* with the imaginary part identified by the letter *j* or *J* (as opposed to the letter *i* used in mathematical notation). For example, the complex number 2 + 3*i* would be written in Python as 2 + 3*j*:

```
>>> a = 2 + 3j
>>> type(a)
<class 'complex'>
```

As you can see, when we use the `type()` function on a complex number, Python tells us that this is an object of type `complex`.

You can also define complex numbers using the `complex()` function:

```
>>> a = complex(2, 3)
>>> a
(2 + 3j)
```

Here we pass the real and imaginary parts of the complex number as two arguments to the `complex()` function, and it returns a complex number.

You can add and subtract complex numbers in the same way as real numbers:

```
>>> b = 3 + 3j
>>> a + b
(5 + 6j)
>>> a - b
(-1 + 0j)
```

Multiplication and division of complex numbers are also carried out similarly:

```
>>> a * b
(-3 + 15j)
>>> a / b
(0.8333333333333334 + 0.16666666666666666j)
```

The modulus (%) and the floor division (//) operations are not valid for complex numbers.

The real and imaginary parts of a complex number can be retrieved using its real and imag attributes, as follows:

```
>>> z = 2 + 3j
>>> z.real
2.0
>>> z.imag
3.0
```

The *conjugate* of a complex number has the same real part but an imaginary part with an equal magnitude and an opposite sign. It can be obtained using the conjugate() method:

```
>>> z.conjugate()
(2 - 3j)
```

Both the real and imaginary parts are floating point numbers. Using the real and imaginary parts, you can then calculate the *magnitude* of a complex number with the following formula, where x and y are the real and imaginary parts of the number, respectively: $\sqrt{x^2 + y^2}$. In Python, this would look like the following:

```
>>> (z.real ** 2 + z.imag ** 2) ** 0.5
3.605551275463989
```

A simpler way to find the magnitude of a complex number is with the abs() function. The abs() function returns the absolute value when called with a real number as its argument. For example, abs(5) and abs(-5) both return 5. However, for complex numbers, it returns the magnitude:

```
>>> abs(z)
3.605551275463989
```

The standard library's cmath module (cmath for *complex math*) provides access to a number of other specialized functions to work with complex numbers.

Getting User Input

As we start to write programs, it will help to have a nice, simple way to accept user input via the input() function. That way, we can write programs that ask a user to input a number, perform specific operations on that number, and then display the results of the operations. Let's see it in action:

```
❶ >>> a = input()
❷ 1
```

At ❶, we call the input() function, which waits for you to type something, as shown at ❷, and press ENTER. The input provided is stored in a:

```
>>> a
❸ '1'
```

Notice the single quotes around 1 at ❸. The input() function returns the input as a *string*. In Python, a string is any set of characters between two quotes. When you want to create a string, either single quotes or double quotes can be used:

```
>>> s1 = 'a string'
>>> s2 = "a string"
```

Here, both s1 and s2 refer to the same string.

Even if the only characters in a string are numbers, Python won't treat that string as a number unless we get rid of those quotation marks. So before we can perform any mathematical operations with the input, we'll have to convert it into the correct number type. A string can be converted to an integer or floating point number using the int() or float() function, respectively:

```
>>> a = '1'
>>> int(a) + 1
2
>>> float(a) + 1
2.0
```

These are the same int() and float() functions we saw earlier, but this time instead of converting the input from one kind of number to another, they take a string as input ('1') and return a number (2 or 2.0). It's important to note, however, that the int() function cannot convert a string containing a floating point decimal into an integer. If you take a string that has a floating point number (like '2.5' or even '2.0') and input that string into the int() function, you'll get an error message:

```
>>> int('2.0')
Traceback (most recent call last):
```

```
    File "<pyshell#26>", line 1, in <module>
      int('2.0')
ValueError: invalid literal for int() with base 10: '2.0'
```

This is an example of an *exception*—Python's way of telling you that it cannot continue executing your program because of an error. In this case, the exception is of the type `ValueError`. (For a quick refresher on exceptions, see Appendix B.)

Similarly, when you supply a fractional number such as 3/4 as an input, Python cannot convert it into an equivalent floating point number or integer. Once again, a `ValueError` exception is raised:

```
>>> a = float(input())
3/4
Traceback (most recent call last):
  File "<pyshell#25>", line 1, in <module>
    a=float(input())
ValueError: could not convert string to float: '3/4'
```

You may find it useful to perform the conversion in a try...except block so that you can *handle* this exception and alert the user that the program has encountered an invalid input. We'll look at try...except blocks next.

Handling Exceptions and Invalid Input

If you're not familiar with try...except, the basic idea is this: if you execute one or more statements in a try...except block and there's an error while executing, your program will not crash and print a Traceback. Instead, the execution is transferred to the except block, where you can perform an appropriate operation, for instance, printing a helpful error message or trying something else.

This is how you would perform the above conversion in a try...except block and print a helpful error message on invalid input:

```
>>> try:
        a = float(input('Enter a number: '))
except ValueError:
        print('You entered an invalid number')
```

Note that we need to specify the type of exception we want to handle. Here, we want to handle the `ValueError` exception, so we specify it as except `ValueError`.

Now, when you give an invalid input, such as 3/4, it prints a helpful error message, as shown at ❶:

```
Enter a number: 3/4
❶ You entered an invalid number
```

You can also specify a prompt with the input() function to tell the user what kind of input is expected. For example:

```
>>> a = input('Input an integer: ')
```

The user will now see the message hinting to enter an integer as input:

```
Input an integer: 1
```

In many programs in this book, we'll ask the user to enter a number as input, so we'll have to make sure we take care of conversion before we attempt to perform any operations on these numbers. You can combine the input and conversion in a single statement, as follows:

```
>>> a = int(input())
1
>>> a + 1
2
```

This works great if the user inputs an integer. But as we saw earlier, if the input is a floating point number (even one that's equivalent to an integer, like 1.0), this will produce an error:

```
>>> a = int(input())
1.0
Traceback (most recent call last):
  File "<pyshell#42>", line 1, in <module>
    a=int(input())
ValueError: invalid literal for int() with base 10: '1.0'
```

In order to avoid this error, we could set up a ValueError catch like the one we saw earlier for fractions. That way the program would catch floating point numbers, which won't work in a program meant for integers. However, it would also flag numbers like 1.0 and 2.0, which Python *sees* as floating point numbers but that are equivalent to integers and would work just fine if they were entered as the right Python type.

To get around all this, we will use the is_integer() method to filter out any numbers with a significant digit after the decimal point. (This method is only defined for float type numbers in Python; it won't work with numbers that are already entered in integer form.)

Here's an example:

```
>>> 1.1.is_integer()
False
```

Here, we call the method is_integer() to check if 1.1 is an integer, and the result is False because 1.1 really is a floating point number. On the other hand, when the method is called with 1.0 as the floating point number, the result is True:

```
>>> 1.0.is_integer()
True
```

We can use is_integer() to filter out noninteger input while keeping inputs like 1.0, which is expressed as a floating point number but is equivalent to an integer. We'll see how the method would fit into a larger program a bit later.

Fractions and Complex Numbers as Input

The Fraction class we learned about earlier is also capable of converting a string such as '3/4' to a Fraction object. In fact, this is how we can accept a fraction as an input:

```
>>> a = Fraction(input('Enter a fraction: '))
Enter a fraction: 3/4
>>> a
Fraction(3, 4)
```

Try entering a fraction such as 3/0 as input:

```
>>> a = Fraction(input('Enter a fraction: '))
Enter a fraction: 3/0
Traceback (most recent call last):
  File "<pyshell#2>", line 1, in <module>
    a = Fraction(input('Enter a fraction: '))
  File "/usr/lib64/python3.3/fractions.py", line 167, in __new__
    raise ZeroDivisionError('Fraction(%s, 0)' % numerator)
ZeroDivisionError: Fraction(3, 0)
```

The ZeroDivisionError exception message tells you (as you already know) that a fraction with a denominator of 0 is invalid. If you're planning on having users enter fractions as input in one of your programs, it's a good idea to always catch such exceptions. Here is how you can do something like that:

```
>>> try:
        a = Fraction(input('Enter a fraction: '))
except ZeroDivisionError:
        print('Invalid fraction')

Enter a fraction: 3/0
Invalid fraction
```

Now, whenever your program's user enters a fraction with 0 in the denominator, it'll print the message Invalid fraction.

Similarly, the complex() function can convert a string such as '2+3j' into a complex number:

```
>>> z = complex(input('Enter a complex number: '))
Enter a complex number: 2+3j
>>> z
(2+3j)
```

If you enter the string as '2 + 3j' (with spaces), it will result in a ValueError error message:

```
>>> z = complex(input('Enter a complex number: '))
Enter a complex number: 2 + 3j
Traceback (most recent call last):
  File "<pyshell#43>", line 1, in <module>
    z = complex(input('Enter a complex number: '))
ValueError: complex() arg is a malformed string
```

It's a good idea to catch the ValueError exception when converting a string to a complex number, as we've done for other number types.

Writing Programs That Do the Math for You

Now that we have learned some of the basic concepts, we can combine them with Python's conditional and looping statements to make some programs that are a little more advanced and useful.

Calculating the Factors of an Integer

When a nonzero integer, a, divides another integer, b, leaving a remainder 0, a is said to be a *factor* of b. As an example, 2 is a factor of all even integers. We can write a function such as the one below to find whether a nonzero integer, a, is a factor of another integer, b:

```
>>> def is_factor(a, b):
        if b % a == 0:
            return True
        else:
            return False
```

We use the % operator introduced earlier in this chapter to calculate the remainder. If you ever find yourself asking a question like "Is 4 a factor of 1024?", you can use the is_factor() function:

```
>>> is_factor(4, 1024)
True
```

For any positive integer n, how do we find all its positive factors? For each of the integers between 1 and n, we check the remainder after dividing n by this integer. If it leaves a remainder of 0, it's a factor. We'll use the range() function to write a program that will go through each of those numbers between 1 and *n*.

Before we write the full program, let's take a look at how range() works. A typical use of the range() function looks like this:

```
>>> for i in range(1, 4):
        print(i)
1
2
3
```

Here, we set up a for loop and gave the range function two arguments. The range() function starts from the integer stated as the first argument (the *start value*) and continues up to the integer just *before* the one stated by the second argument (the *stop value*). In this case, we told Python to print out the numbers in that range, beginning with 1 and stopping at 4. Note that this means Python doesn't print 4, so the last number it prints is the number before the stop value (3). It's also important to note that the range() function accepts only integers as its arguments.

You can also use the range() function without specifying the start value, in which case it's assumed to be 0. For example:

```
>>> for i in range(5):
        print(i)
0
1
2
3
4
```

The difference between two consecutive integers produced by the range() function is known as the *step value*. By default, the step value is 1. To specify a different step value, specify it as the third argument (the start value is *not* optional when you specify a step value). For example, the following program prints the odd numbers *below* 10:

```
>>> for i in range(1,10,2):
        print(i)
1
3
5
7
9
```

Okay, now that we see how the range() function works, we're ready to look at a factor-calculating program. Because I'm writing a fairly long program, instead of writing this program in the interactive IDLE prompt, I write it in the IDLE editor. You can start the editor by selecting **File ▸ New Window** in IDLE. Notice that we start out by commenting our code with three straight single quotes ('). The text in between those quotes won't be executed by Python as part of the program; it's just commentary for us humans.

```
'''
Find the factors of an integer
'''

def factors(b):
❶    for i in range(1, b+1):
        if b % i == 0:
            print(i)

if __name__ == '__main__':

    b = input('Your Number Please: ')
    b = float(b)

❷    if b > 0 and b.is_integer():
        factors(int(b))
    else:
        print('Please enter a positive integer')
```

The factors() function defines a for loop that iterates once for every integer between 1 and the input integer at ❶ using the range() function. Here, we want to iterate up to the integer entered by the user, b, so the stop value is stated as b+1. For each of these integers, i, the program checks whether it divides the input number with no remainder and prints it if so.

When you run this program (by selecting **Run ▸ Run Module**), it asks you to input a number. If your number is a positive integer, its factors are printed. For example:

```
Your Number Please: 25
1
5
25
```

If you enter a non-integer or a negative integer as an input, the program prints an error message asking you to input a positive integer:

```
Your Number Please: 15.5
Please enter a positive integer
```

This is an example of how we can make programs more user friendly by always checking for invalid input in the program itself. Because our program works only for finding the factors of a positive integer, we check whether the input number is greater than 0 and is an integer using the is_integer() method ❷ to make sure the input is valid. If the input isn't a positive integer, the program prints a user-friendly instruction instead of just a big error message.

Generating Multiplication Tables

Consider three numbers, *a*, *b*, and *n*, where *n* is an integer, such that

$$a \times n = b.$$

We can say here that *b* is the *n*th *multiple* of *a*. For example, 4 is the 2nd multiple of 2, and 1024 is the 512nd multiple of 2.

A multiplication table for a number lists all of that number's multiples. For example, the multiplication table of 2 looks like this (first three multiples shown here):

$$2 \times 1 = 2$$

$$2 \times 2 = 4$$

$$2 \times 3 = 6$$

Our next program generates the multiplication number up to 10 for any number input by the user. In this program, we'll use the `format()` method with the `print()` function to help make the program's output look nicer and more readable. In case you haven't seen it before, I'll now briefly explain how it works.

The `format()` method lets you plug in labels and set it up so that they get printed out in a nice, readable string with extra formatting around it. For example, if I had the names of all the fruits I bought at the grocery store with separate labels created for each and wanted to print them out to make a coherent sentence, I could use the `format()` method as follows:

```
>>> item1 = 'apples'
>>> item2 = 'bananas'
>>> item3 = 'grapes'
>>> print('At the grocery store, I bought some {0} and {1} and {2}'.format(item1, item2, item3))
At the grocery store, I bought some apples and bananas and grapes
```

First, we created three labels (`item1`, `item2`, and `item3`), each referring to a different string (apples, bananas, and grapes). Then, in the `print()` function, we typed a string with three placeholders in curly brackets: {0}, {1}, and {2}. We followed this with `.format()`, which holds the three labels we created. This tells Python to fill those three placeholders with the values stored in those labels in the order listed, so Python prints the text with {0} replaced by the first label, {1} replaced by the second label, and so on.

It's not necessary to have labels pointing to the values we want to print. We can also just type values into `.format()`, as in the following example:

```
>>> print('Number 1: {0} Number 2: {1} '.format(1, 3.578))
Number 1: 1 Number 2: 3.578
```

Note that the number of placeholders and the number of labels or values must be equal.

Now that we've seen how `format()` works, we're ready to take a look at the program for our multiplication table printer:

```
'''
Multiplication table printer
'''

def multi_table(a):

❶    for i in range(1, 11):
        print('{0} x {1} = {2}'.format(a, i, a*i))


if __name__ == '__main__':
    a = input('Enter a number: ')
    multi_table(float(a))
```

The function `multi_table()` implements the main functionality of the program. It takes the number for which the multiplication table will be printed as a parameter, a. Because we want to print the multiplication table from 1 to 10, we have a for loop at ❶ that iterates over each of these numbers, printing the product of itself and the number, a.

When you execute the program, it asks you to input a number, and the program prints its multiplication table:

```
Enter a number : 5
5.0 x 1 = 5.0
5.0 x 2 = 10.0
5.0 x 3 = 15.0
5.0 x 4 = 20.0
5.0 x 5 = 25.0
5.0 x 6 = 30.0
5.0 x 7 = 35.0
5.0 x 8 = 40.0
5.0 x 9 = 45.0
5.0 x 10 = 50.0
```

See how nice and orderly that table looks? That's because we used the `.format()` method to print the output according to a readable, uniform template.

You can use the `format()` method to further control how numbers are printed. For example, if you want numbers with only two decimal places, you can specify that with the `format()` method. Here is an example:

```
>>> '{0}'.format(1.25456)
'1.25456'
>>> '{0:.2f}'.format(1.25456)
'1.25'
```

The first format statement above simply prints the number exactly as we entered it. In the second statement, we modify the place holder to {0:.2f}, meaning that we want only two numbers after the decimal point, with the f indicating a floating point number. As you can see, there are only two numbers after the decimal point in the next output. Note that the number is rounded if there are more numbers after the decimal point than you specified. For example:

```
>>> '{0:.2f}'.format(1.25556)
'1.26'
```

Here, 1.25556 is rounded up to the nearest hundredth and printed as 1.26. If you use .2f and the number you are printing is an integer, zeros are added at the end:

```
>>> '{0:.2f}'.format(1)
'1.00'
```

Two zeros are added because we specified that we should print exactly two numbers after the decimal point.

Converting Units of Measurement

The International System of Units defines seven *base quantities*. These are then used to derive other quantities, referred to as *derived quantities*. Length (including width, height, and depth), time, mass, and temperature are four of the seven base quantities. Each of these quantities has a standard unit of measurement: meter, second, kilogram, and kelvin, respectively.

But each of these standard measurement units also has multiple nonstandard measurement units. You are more familiar with the temperature being reported as 30 degrees Celsius or 86 degrees Fahrenheit than as 303.15 kelvin. Does that mean 303.15 kelvin feels three times hotter than 86 degrees Fahrenheit? No way! We can't compare 86 degrees Fahrenheit to 303.15 kelvin only by their numerical values because they're expressed in different measurement units, even though they measure the same physical quantity—temperature. You can compare two measurements of a physical quantity only when they're expressed in the same unit of measurement.

Conversions between different units of measurement can be tricky, and that's why you're often asked to solve problems that involve conversion between different units of measurement in high school. It's a good way to test your basic mathematical skills. But Python has plenty of math skills, too, and, unlike some high school students, it doesn't get tired of crunching numbers over and over again in a loop! Next, we'll explore writing programs to perform those unit conversions for you.

We'll start with length. In the United States and United Kingdom, inches and miles are often used for measuring length, while most other countries use centimeters and kilometers.

An inch is equal to 2.54 centimeters, and you can use the multiplication operation to convert a measurement in inches to centimeters. You can then divide the measurement in centimeters by 100 to obtain the measurement in meters. For example, here's how you can convert 25.5 inches to meters:

```
>>> (25.5 * 2.54) / 100
0.6476999999999999
```

On the other hand, a mile is roughly equivalent to 1.609 kilometers. So if you see that your destination is 650 miles away, you're 650 × 1.609 kilometers away:

```
>>> 650 * 1.609
1045.85
```

Now let's take a look at *temperature* conversion—converting temperature from Fahrenheit to Celsius and vice versa. Temperature expressed in Fahrenheit is converted into its equivalent value in Celsius using the formula

$$C = \left(F - 32\right) \times \frac{5}{9}.$$

F is the temperature in Fahrenheit, and *C* is its equivalent in Celsius. You know that 98.6 degrees Fahrenheit is said to be the normal human body temperature. To find the corresponding temperature in degrees Celsius, we evaluate the above formula in Python:

```
>>> F = 98.6
>>> (F - 32) * (5 / 9)
37.0
```

First, we create a label, F, with the temperature in Fahrenheit, 98.6. Next, we evaluate the formula for converting this temperature to its equivalent in Celsius, which turns out be 37.0 degrees Celsius.

To convert temperature from Celsius to Fahrenheit, you would use the formula

$$F = \left(C \times \frac{9}{5}\right) + 32.$$

You can evaluate this formula in a similar manner:

```
>>> C = 37
>>> C * (9 / 5) + 32
98.60000000000001
```

We create a label, C, with the value 37 (the normal human body temperature in Celsius). Then, we convert it into Fahrenheit using the formula, and the result is 98.6 degrees.

It's a chore to have to write these conversion formulas over and over again. Let's write a unit conversion program that will do the conversions for us. This program will present a menu to allow users to select the conversion they want to perform, ask for relevant input, and then print the calculated result. The program is shown below:

```
'''
Unit converter: Miles and Kilometers
'''

def print_menu():
    print('1. Kilometers to Miles')
    print('2. Miles to Kilometers')

def km_miles():
    km = float(input('Enter distance in kilometers: '))
    miles = km / 1.609

    print('Distance in miles: {0}'.format(miles))

def miles_km():
    miles = float(input('Enter distance in miles: '))
    km = miles * 1.609

    print('Distance in kilometers: {0}'.format(km))

if __name__ == '__main__':
❶    print_menu()
❷    choice = input('Which conversion would you like to do?: ')
    if choice == '1':
        km_miles()

    if choice == '2':
        miles_km()
```

This is a slightly longer program than the others, but not to worry. It's actually simple. Let's start from ❶. The print_menu() function is called, which prints a menu with two unit conversion choices. At ❷, the user is asked to select one of the two conversions. If the choice is entered as 1 (kilometers to miles), the function km_miles() is called. If the choice is entered as 2 (miles to kilometers), the function miles_km() is called. In both of these functions, the user is first asked to enter a distance in the unit chosen for conversion (kilometers for km_miles() and miles for miles_km()). The program then performs the conversion using the corresponding formula and displays the result.

Here is a sample run of the program:

```
1. Kilometers to Miles
2. Miles to Kilometers
❶ Which conversion would you like to do?: 2
Enter distance in miles: 100
Distance in kilometers: 160.900000
```

The user is asked to enter a choice at ❶. The choice is entered as 2 (miles to kilometers). The program then asks the user to enter the distance in miles to be converted to kilometers and prints the conversion.

This program just converts between miles and kilometers, but in a programming challenge later, you'll extend this program so that it can perform conversions of other units.

Finding the Roots of a Quadratic Equation

What do you do when you have an equation such as $x + 500 − 79 = 10$ and you need to find the value of the unknown variable, x? You rearrange the terms such that you have only the constants (500, −79, and 10) on one side of the equation and the variable (x) on the other side. This results in the following equation: $x = 10 − 500 + 79$.

Finding the value of the expression on the right gives you the value of x, your solution, which is also called the *root* of this equation. In Python, you can do this as follows:

```
>>> x = 10 - 500 + 79
>>> x
-411
```

This is an example of a *linear equation*. Once you have rearranged the terms on both sides, the expression is simple enough to evaluate. On the other hand, for equations such as $x^2 + 2x + 1 = 0$, finding the roots of x usually involves evaluating a complex expression known as the *quadratic formula*. Such equations are known as *quadratic equations*, generally expressed as $ax^2 + bx + c = 0$, where a, b, and c are constants. The quadratic formula for calculating the roots is given as follows:

$$x_1 = \frac{-b + \sqrt{b^2 - 4ac}}{2a} \quad \text{and} \quad x_2 = \frac{-b - \sqrt{b^2 - 4ac}}{2a}.$$

A quadratic equation has two roots—two values of x for which the two sides of the quadratic equation are equal (although sometimes these two values may turn out to be the same). This is indicated here by the x_1 and x_2 in the quadratic formula.

Comparing the equation $x^2 + 2x + 1 = 0$ to the generic quadratic equation, we see that $a = 1$, $b = 2$, and $c = 1$. We can substitute these values directly into the quadratic formula to calculate the value of x_1 and x_2. In Python, we first store the values of a, b, and c as the labels a, b, and c with the appropriate values:

```
>>> a = 1
>>> b = 2
>>> c = 1
```

Then, considering that both the formulas have the term $b^2 - 4ac$, we'll define a new label with D, such that $D = \sqrt{b^2 - 4ac}$:

```
>>> D = (b**2 - 4*a*c)**0.5
```

As you can see, we evaluate the square root of $b^2 - 4ac$ by raising it to the 0.5th power. Now, we can write the expressions for evaluating x_1 and x_2:

```
>>> x_1 = (-b + D)/(2*a)
>>> x_1
-1.0
>>> x_2 = (-b - D)/(2*a)
>>> x_2
-1.0
```

In this case, the values of both the roots are the same, and if you substitute that value into the equation $x^2 + 2x + 1$, the equation will evaluate to 0.

Our next program combines all these steps in a function roots(), which takes the values of a, b, and c as parameters, calculates the roots, and prints them:

```
'''
Quadratic equation root calculator
'''

def roots(a, b, c):

    D = (b*b - 4*a*c)**0.5
    x_1 = (-b + D)/(2*a)
    x_2 = (-b - D)/(2*a)

    print('x1: {0}'.format(x_1))
    print('x2: {0}'.format(x_2))

if __name__ == '__main__':
    a = input('Enter a: ')
    b = input('Enter b: ')
    c = input('Enter c: ')
    roots(float(a), float(b), float(c))
```

At first, we use the labels a, b, and c to reference the values of the three constants of a quadratic equation. Then, we call the roots() function with these three values as arguments (after converting them to floating point numbers). This function plugs a, b, and c into the quadratic formula, finds the roots for that equation, and prints them.

When you execute the program, it will ask the user to input values of a, b, and c corresponding to a quadratic equation they want to find the roots for.

```
Enter a: 1
Enter b: 2
Enter c: 1
```

```
x1: -1.000000
x2: -1.000000
```

Try solving a few more quadratic equations with different values for the constants, and the program will find the roots correctly.

You most likely know that quadratic equations can have complex numbers as roots, too. For example, the roots of the equation $x^2 + x + 1 = 0$ are both complex numbers. The above program can find those for you as well. Let's give it a shot by executing the program again (the constants are $a = 1$, $b = 1$, and $c = 1$):

```
Enter a: 1
Enter b: 1
Enter c: 1
x1: (-0.49999999999999994+0.8660254037844386j)
x2: (-0.5-0.8660254037844386j)
```

The roots printed above are complex numbers (indicated by j), and the program has no problem calculating or displaying them.

What You Learned

Great work on finishing the first chapter! You learned to write programs that recognize integers, floating point numbers, fractional numbers (expressed as a fraction or a floating point number), and complex numbers. You wrote programs that generate multiplication tables, perform unit conversions, and find the roots of a quadratic equation. I'm sure you're already excited about having taken the first steps toward writing programs that will do mathematical calculations for you. Before we move on, here are some programming challenges that will give you a chance to further apply what you've learned.

Programming Challenges

Here are a few challenges that will give you a chance to practice the concepts from this chapter. Each problem can be solved in multiple ways, but you can find sample solutions at *http://www.nostarch.com/doingmathwithpython/*.

#1: Even-Odd Vending Machine

Try writing an "even-odd vending machine," which will take a number as input and do two things:

1. Print whether the number is even or odd.
2. Display the number followed by the next 9 even or odd numbers.

If the input is 2, the program should print even and then print 2, 4, 6, 8, 10, 12, 14, 16, 18, 20. Similarly, if the input is 1, the program should print odd and then print 1, 3, 5, 7, 9, 11, 13, 15, 17, 19.

Your program should use the is_integer() method to display an error message if the input is a number with significant digits beyond the decimal point.

#2: Enhanced Multiplication Table Generator

Our multiplication table generator is cool, but it prints only the first 10 multiples. Enhance the generator so that the user can specify both the number and up to *which* multiple. For example, I should be able to input that I want to see a table listing the first 15 multiples of 9.

#3: Enhanced Unit Converter

The unit conversion program we wrote in this chapter is limited to conversions between kilometers and miles. Try extending the program to convert between units of mass (such as kilograms and pounds) and between units of temperature (such as Celsius and Fahrenheit).

#4: Fraction Calculator

Write a calculator that can perform the basic mathematical operations on two fractions. It should ask the user for two fractions and the operation the user wants to carry out. As a head start, here's how you can write the program with only the addition operation:

```
'''
Fraction operations
'''
from fractions import Fraction

def add(a, b):
    print('Result of Addition: {0}'.format(a+b))

if __name__ == '__main__':
❶    a = Fraction(input('Enter first fraction: '))
❷    b = Fraction(input('Enter second fraction: '))
    op = input('Operation to perform - Add, Subtract, Divide, Multiply: ')
    if op == 'Add':
        add(a,b)
```

You've already seen most of the elements in this program. At ❶ and ❷, we ask the user to input the two fractions. Then, we ask the user which operation is to be performed on the two fractions. If the user enters 'Add' as input, we call the function add(), which we've defined to find the sum of the two fractions passed as arguments. The add() function performs the operation and prints the result. For example:

```
Enter first fraction: 3/4
Enter second fraction: 1/4
Operation to perform - Add, Subtract, Divide, Multiply: Add
Result of Addition: 1
```

Try adding support for other operations such as subtraction, division, and multiplication. For example, here's how your program should be able to calculate the difference of two fractions:

```
Enter first fraction: 3/4
Enter second fraction: 1/4
Operation to perform - Add, Subtract, Divide, Multiply: Subtract
Result of Subtraction: 2/4
```

In the case of division, you should let the user know whether the first fraction is divided by the second fraction or vice versa.

#5: Give Exit Power to the User

All the programs we have written so far work only for one iteration of input and output. For example, consider the program to print the multiplication table: the user executes the program and enters a number; then the program prints the multiplication table and exits. If the user wanted to print the multiplication table of another number, the program would have to be rerun.

It would be more convenient if the user could choose whether to exit or continue using the program. The key to writing such programs is to set up an *infinite loop*, or a loop that doesn't exit unless explicitly asked to do so. Below, you can see an example of the layout for such a program:

```
'''
Run until exit layout
'''

def fun():
    print('I am in an endless loop')

if __name__ == '__main__':
❶    while True:
        fun()
❷        answer = input('Do you want to exit? (y) for yes ')
        if answer == 'y':
            break
```

We define an infinite loop using while True at ❶. A while loop continues to execute unless the condition evaluates to False. Because we chose the loop's condition to be the constant value True, it will keep running forever unless we interrupt it somehow. Inside the loop, we call the function fun(), which prints the string I am in an endless loop. At ❷, the user is asked "Do you want to exit?" If the user enters y as the input, the program exits out of the loop using the break statement (break exits out of the innermost loop without executing any other statement in that loop). If the user enters any other input (or none at all, just pressing ENTER), the while loop continues execution—that is, it prints the string again and continues doing so until the user wishes to exit. Here is a sample run of the program:

```
I am in an endless loop
Do you want to exit? (y) for yes n
I am in an endless loop
Do you want to exit? (y) for yes n
I am in an endless loop
Do you want to exit? (y) for yes n
I am in an endless loop
Do you want to exit? (y) for yes y
```

Based on this example, let's rewrite the multiplication table generator so that it keeps going until the user wants to exit. The new version of the program is shown below:

```
'''
Multiplication table printer with
exit power to the user
'''

def multi_table(a):

    for i in range(1, 11):
        print('{0} x {1} = {2}'.format(a, i, a*i))

if __name__ == '__main__':

    while True:
        a = input('Enter a number: ')
        multi_table(float(a))

        answer = input('Do you want to exit? (y) for yes ')
        if answer == 'y':
            break
```

If you compare this program to the one we wrote earlier, you'll see that the only change is the addition of the while loop, which includes the prompt asking the user to input a number and the call to the multi_table() function.

When you run the program, the program will ask for a number and print its multiplication table, as before. However, it will also subsequently ask whether the user wants to exit the program. If the user doesn't want to exit, the program will be ready to print the table for another number. Here is a sample run:

```
Enter a number: 2
2.000000 x 1.000000 = 2.000000
2.000000 x 2.000000 = 4.000000
2.000000 x 3.000000 = 6.000000
2.000000 x 4.000000 = 8.000000
```

```
2.000000 x 5.000000 = 10.000000
2.000000 x 6.000000 = 12.000000
2.000000 x 7.000000 = 14.000000
2.000000 x 8.000000 = 16.000000
2.000000 x 9.000000 = 18.000000
2.000000 x 10.000000 = 20.000000
Do you want to exit? (y) for yes n
Enter a number:
```

Try rewriting some of the other programs in this chapter so that they continue executing until asked by the user to exit.

2

VISUALIZING DATA WITH GRAPHS

In this chapter, you'll learn a powerful way to present numerical data: by drawing graphs with Python. We'll start by discussing the number line and the Cartesian plane. Next, we'll learn about the powerful plotting library *matplotlib* and how we can use it to create graphs. We'll then explore how to make graphs that present data clearly and intuitively. Finally, we'll use graphs to explore Newton's law of universal gravitation and projectile motion. Let's get started!

Understanding the Cartesian Coordinate Plane

Consider a *number line*, like the one shown in Figure 2-1. Integers from −3 to 3 are marked on the line, but between any of these two numbers (say, 1 and 2) lie all possible numbers in between: 1.1, 1.2, 1.3, and so on.

Figure 2-1: A number line

The number line makes certain properties visually intuitive. For example, all numbers on the right side of 0 are positive, and those on the left side are negative. When a number *a* lies on the right side of another number *b*, *a* is always greater than *b* and *b* is always less than *a*.

The arrows at the ends of the number line indicate that the line extends infinitely, and any point on this line corresponds to some real number, however large it may be. A single number is sufficient to describe a point on the number line.

Now consider two number lines arranged as shown in Figure 2-2. The number lines intersect at right angles to each other and cross at the 0 point of each line. This forms a *Cartesian coordinate plane*, or an *x-y* plane, with the horizontal number line called the *x*-axis and the vertical line called the *y*-axis.

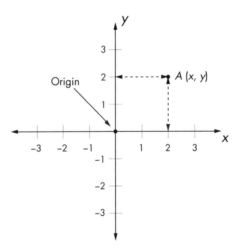

Figure 2-2: The Cartesian coordinate plane

As with the number line, we can have infinitely many points on the plane. We describe a point with a pair of numbers instead of one number. For example, we describe the point *A* in the figure with two numbers, *x* and *y*, usually written as (*x*, *y*) and referred to as the *coordinates* of the point.

As shown in Figure 2-2, *x* is the distance of the point from the origin along the *x*-axis, and *y* is the distance along the *y*-axis. The point where the two axes intersect is called the *origin* and has the coordinates (0, 0).

The Cartesian coordinate plane allows us to visualize the relationship between two sets of numbers. Here, I use the term *set* loosely to mean a collection of numbers. (We'll learn about mathematical sets and how to work with them in Python in Chapter 5.) No matter what the two sets of numbers represent—temperature, baseball scores, or class test scores—all you need are the numbers themselves. Then, you can plot them—either on graph paper or on your computer with a program written in Python. For the rest of this book, I'll use the term *plot* as a verb to describe the act of plotting two sets of numbers and the term *graph* to describe the result—a line, curve, or simply a set of points on the Cartesian plane.

Working with Lists and Tuples

As we make graphs with Python, we'll work with *lists* and *tuples*. In Python, these are two different ways to store groups of values. Tuples and lists are very similar for the most part, with one major difference: after you create a list, it's possible to add values to it and to change the order of the values. The values in a tuple, on the other hand, are immediately fixed and can't be changed. We'll use lists to store *x*- and *y*-coordinates for the points we want to plot. Tuples will come up in "Customizing Graphs" on page 41 when we learn to customize the range of our graphs. First, let's go over some features of lists.

You can create a list by entering values, separated by commas, between square brackets. The following statement creates a list and uses the label simplelist to refer to it:

```
>>> simplelist = [1, 2, 3]
```

Now you can refer to the individual numbers—1, 2, and 3—using the label and the position of the number in the list, which is called the *index.* So simplelist[0] refers to the first number, simplelist[1] refers to the second number, and simplelist[2] refers to the third number:

```
>>> simplelist[0]
1
>>> simplelist[1]
2
>>> simplelist[2]
3
```

Notice that the first item of the list is at index 0, the second item is at index 1, and so on—that is, the positions in the list start counting from 0, not 1.

Lists can store strings, too:

```
>>> stringlist = ['a string','b string','c string']
>>> stringlist[0]
'a string'
>>> stringlist[1]
'b string'
>>> stringlist[2]
'c string'
```

One advantage of creating a list is that you don't have to create a separate label for each value; you just create a label for the list and use the index position to refer to each item. Also, you can add to the list whenever you need to store new values, so a list is the best choice for storing data if you don't know beforehand how many numbers or strings you may need to store.

An *empty list* is just that—a list with no items or elements—and it can be created like this:

```
>>> emptylist = []
```

Empty lists are mainly useful when you don't know any of the items that will be in your list beforehand but plan to fill in values during the execution of a program. In that case, you can create an empty list and then use the append() method to add items later:

```
❶ >>> emptylist
   []
❷ >>> emptylist.append(1)
   >>> emptylist
   [1]
❸ >>> emptylist.append(2)
   >>> emptylist
❹ [1, 2]
```

At ❶, emptylist starts off empty. Next, we append the number 1 to the list at ❷ and then append 2 at ❸. By line ❹, the list is now [1, 2]. Note that when you use .append(), the value gets added to the end of the list. This is just one way of adding values to a list. There are others, but we won't need them for this chapter.

Creating a tuple is similar to creating a list, but instead of square brackets, you use parentheses:

```
>>> simpletuple = (1, 2, 3)
```

You can refer to an individual number in simpletuple using the corresponding index in brackets, just as with lists:

```
>>> simpletuple[0]
1
```

```
>>> simpletuple[1]
2
>>> simpletuple[2]
3
```

You can also use *negative indices* with both lists and tuples. For example, simplelist[-1] and simpletuple[-1] would refer to the last element of the list or the tuple, simplelist[-2] and simpletuple[-2] would refer to the second-to-last element, and so on.

Tuples, like lists, can have strings as values, and you can create an *empty tuple* with no elements as emptytuple=(). However, there's no append() method to add a new value to an existing tuple, so you can't add values to an empty tuple. Once you create a tuple, the contents of the tuple can't be changed.

Iterating over a List or Tuple

We can go over a list or tuple using a for loop as follows:

```
>>> l = [1, 2, 3]
>>> for item in l:
        print(item)
```

This will print the items in the list:

```
1
2
3
```

The items in a tuple can be retrieved in the same way.

Sometimes you might need to know the position or the index of an item in a list or tuple. You can use the enumerate() function to iterate over all the items of a list and return the index of an item as well as the item itself. We use the labels index and item to refer to them:

```
>>> l = [1, 2, 3]
>>> for index, item in enumerate(l):
        print(index, item)
```

This will produce the following output:

```
0 1
1 2
2 3
```

This also works for tuples.

Creating Graphs with Matplotlib

We'll be using matplotlib to make graphs with Python. Matplotlib is a Python *package*, which means that it's a collection of modules with related functionality. In this case, the modules are useful for plotting numbers and making graphs. Matplotlib doesn't come built in with Python's standard library, so you'll have to install it. The installation instructions are covered in Appendix A. Once you have it installed, start a Python shell. As explained in the installation instructions, you can either continue using IDLE shell or use Python's built-in shell.

Now we're ready to create our first graph. We'll start with a simple graph with just three points: (1, 2), (2, 4), and (3, 6). To create this graph, we'll first make two lists of numbers—one storing the values of the *x*-coordinates of these points and another storing the *y*-coordinates. The following two statements do exactly that, creating the two lists x_numbers and y_numbers:

```
>>> x_numbers = [1, 2, 3]
>>> y_numbers = [2, 4, 6]
```

From here, we can create the plot:

```
>>> from pylab import plot, show
>>> plot(x_numbers, y_numbers)
[<matplotlib.lines.Line2D object at 0x7f83ac60df10>]
```

In the first line, we import the plot() and show() functions from the pylab module, which is part of the matplotlib package. Next, we call the plot() function in the second line. The first argument to the plot() function is the list of numbers we want to plot on the *x*-axis, and the second argument is the corresponding list of numbers we want to plot on the *y*-axis. The plot() function returns an object—or more precisely, a list containing an object. This object contains the information about the graph that we asked Python to create. At this stage, you can add more information, such as a title, to the graph, or you can just display the graph as it is. For now we'll just display the graph.

The plot() function only creates the graph. To actually display it, we have to call the show() function:

```
>>> show()
```

You should see the graph in a matplotlib window as shown in Figure 2-3. (The display window may look different depending on your operating system, but the graph should be the same.)

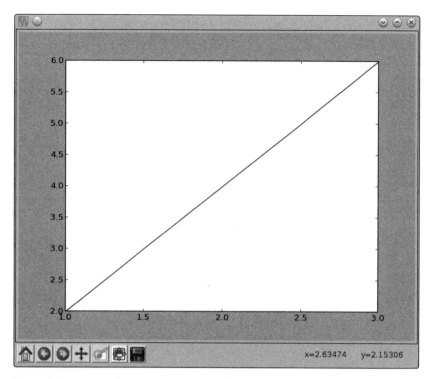

Figure 2-3: A graph showing a line passing through the points (1, 2), (2, 4), and (3, 6)

Notice that instead of starting from the origin (0, 0), the *x*-axis starts from the number 1 and the *y*-axis starts from the number 2. These are the lowest numbers from each of the two lists. Also, you can see increments marked on each of the axes (such as 2.5, 3.0, 3.5, etc., on the *y*-axis). In "Customizing Graphs" on page 41, we'll learn how to control those aspects of the graph, along with how to add axes labels and a graph title.

You'll notice in the interactive shell that you can't enter any further statements until you close the matplotlib window. Close the graph window so that you can continue programming.

Marking Points on Your Graph

If you want the graph to mark the points that you supplied for plotting, you can use an additional keyword argument while calling the plot() function:

```
>>> plot(x_numbers, y_numbers, marker='o')
```

By entering `marker='o'`, we tell Python to mark each point from our lists with a small dot that looks like an *o*. Once you enter `show()` again, you'll see that each point is marked with a dot (see Figure 2-4).

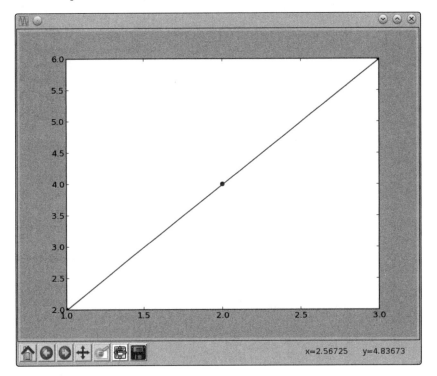

Figure 2-4: A graph showing a line passing through the points (1, 2), (2, 4), and (3, 6) with the points marked by a dot

The marker at (2, 4) is easily visible, while the others are hidden in the very corners of the graph. You can choose from several `marker` options, including `'o'`, `'*'`, `'x'`, and `'+'`. Using `marker=` includes a line connecting the points (this is the default). You can also make a graph that marks only the points that you specified, without any line connecting them, by omitting `marker=`:

```
>>> plot(x_numbers, y_numbers, 'o')
[<matplotlib.lines.Line2D object at 0x7f2549bc0bd0>]
```

Here, `'o'` indicates that each point should be marked with a dot, but there should be no line connecting the points. Call the function `show()` to display the graph, which should look like the one shown in Figure 2-5.

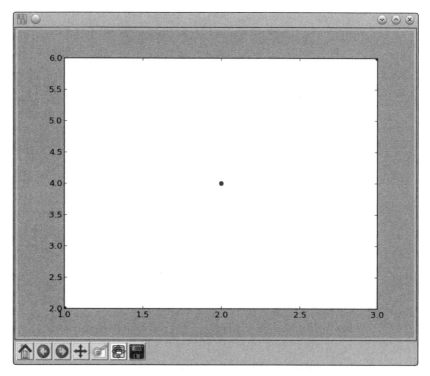

Figure 2-5: A graph showing the points (1, 2), (2, 4), and (3, 6)

As you can see, only the points are now shown on the graph, with no line connecting them. As in the previous graph, the first and the last points are barely visible, but we'll soon see how to change that.

Graphing the Average Annual Temperature in New York City

Let's take a look at a slightly larger set of data so we can explore more features of matplotlib. The average annual temperatures for New York City—measured at Central Park, specifically—during the years 2000 to 2012 are as follows: 53.9, 56.3, 56.4, 53.4, 54.5, 55.8, 56.8, 55.0, 55.3, 54.0, 56.7, 56.4, and 57.3 degrees Fahrenheit. Right now, that just looks like a random jumble of numbers, but we can plot this set of temperatures on a graph to make the rise and fall in the average temperature from year to year much clearer:

```
>>> nyc_temp = [53.9, 56.3, 56.4, 53.4, 54.5, 55.8, 56.8, 55.0, 55.3, 54.0, 56.7, 56.4, 57.3]
>>> plot(nyc_temp, marker='o')
[<matplotlib.lines.Line2D object at 0x7f2549d52f90>]
```

We store the average temperatures in a list, nyc_temp. Then, we call the function plot() passing only this list (and the marker string). When you use plot() on a single list, those numbers are automatically plotted on the *y*-axis. The corresponding values on the *x*-axis are filled in as the positions of each value in the list. That is, the first temperature value, 53.9, gets a corresponding *x*-axis value of 0 because it's in position 0 of the list (remember, the list position starts counting from 0, not 1). As a result, the numbers plotted on the *x*-axis are the integers from 0 to 12, which we can think of as corresponding to the 13 years for which we have temperature data.

Enter show() to display the graph, which is shown in Figure 2-6. The graph shows that the average temperature has risen and fallen from year to year. If you glance at the numbers we plotted, they really aren't very far apart from each other. However, the graph makes the variations seem rather dramatic. So, what's going on? The reason is that matplotlib chooses the range of the *y*-axis so that it's just enough to enclose the data supplied for plotting. So in this graph, the *y*-axis starts at 53.0 and its highest value is 57.5. This makes even small differences look magnified because the range of the *y*-axis is so small. We'll learn how to control the range of each axis in "Customizing Graphs" on page 41.

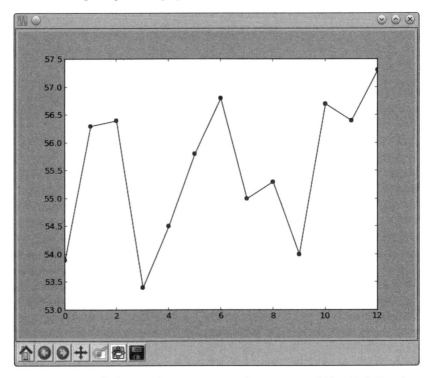

Figure 2-6: A graph showing the average annual temperature of New York City during the years 2000–2012

You can also see that numbers on the *y*-axis are floating point numbers (because that's what we asked to be plotted) and those on the *x*-axis are integers. Matplotlib can handle either.

Plotting the temperature without showing the corresponding years is a quick and easy way to visualize the variations between the years. If you were planning to present this graph to someone, however, you'd want to make it clearer by showing which year each temperature corresponds to. We can easily do this by creating another list with the years in it and then calling the plot() function:

```
>>> nyc_temp = [53.9, 56.3, 56.4, 53.4, 54.5, 55.8, 56.8, 55.0, 55.3, 54.0, 56.7, 56.4, 57.3]
>>> years = range(2000, 2013)
>>> plot(years, nyc_temp, marker='o')
[<matplotlib.lines.Line2D object at 0x7f2549a616d0>]
>>> show()
```

We use the range() function we learned about in Chapter 1 to specify the years 2000 to 2012. Now you'll see the years displayed on the *x*-axis (see Figure 2-7).

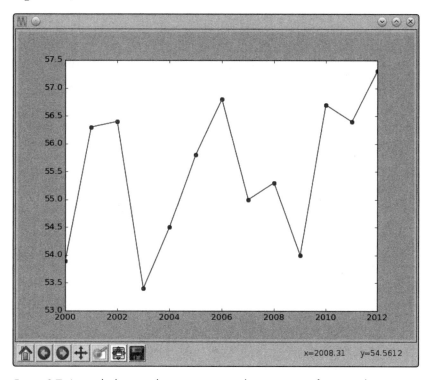

Figure 2-7: A graph showing the average annual temperature of New York City, displaying the years on the x-axis

Comparing the Monthly Temperature Trends of New York City

While still looking at New York City, let's see how the average monthly temperature has varied over the years. This will give us a chance to understand how to plot multiple lines on a single graph. We'll choose three years: 2000, 2006, and 2012. For each of these years, we'll plot the average temperature for all 12 months.

First, we need to create three lists to store the temperature (in Fahrenheit). Each list will consist of 12 numbers corresponding to the average temperature from January to December each year:

```
>>> nyc_temp_2000 = [31.3, 37.3, 47.2, 51.0, 63.5, 71.3, 72.3, 72.7, 66.0, 57.0, 45.3, 31.1]
>>> nyc_temp_2006 = [40.9, 35.7, 43.1, 55.7, 63.1, 71.0, 77.9, 75.8, 66.6, 56.2, 51.9, 43.6]
>>> nyc_temp_2012 = [37.3, 40.9, 50.9, 54.8, 65.1, 71.0, 78.8, 76.7, 68.8, 58.0, 43.9, 41.5]
```

The first list corresponds to the year 2000, and the next two lists correspond to the years 2006 and 2012, respectively. We could plot the three sets of data on three different graphs, but that wouldn't make it very easy to see how each year compares to the others. Try doing it!

The clearest way to compare all of these temperatures is to plot all three data sets on a *single* graph, like this:

```
>>> months = range(1, 13)
>>> plot(months, nyc_temp_2000, months, nyc_temp_2006, months, nyc_temp_2012)
[<matplotlib.lines.Line2D object at 0x7f2549c1f0d0>, <matplotlib.lines.Line2D
object at 0x7f2549a61150>, <matplotlib.lines.Line2D object at 0x7f2549c1b550>]
```

First, we create a list (months) where we store the numbers 1, 2, 3, and so on up to 12 using the range() function. Next, we call the plot() function with three pairs of lists. Each pair consists of a list of months to be plotted on the *x*-axis and a list of average monthly temperatures (for 2000, 2006, and 2012, respectively) to be plotted on the *y*-axis. So far, we've used plot() on only one pair of lists at a time, but you can actually enter multiple pairs of lists into the plot() function. With each list separated by a comma, the plot() function will automatically plot a different line for each pair.

The plot() function returns a list of three objects instead of one. Matplotlib considers the three curves as distinct from each other, and it knows to draw them on top of each other when you call show(). Let's call show() to display the graph, as shown in Figure 2-8.

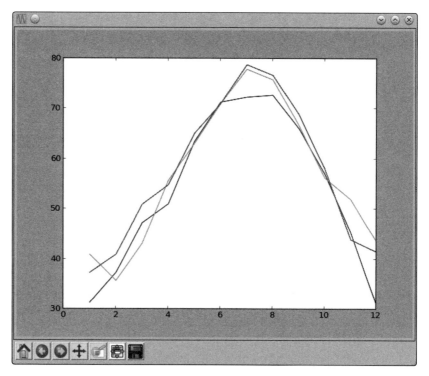

Figure 2-8: A graph showing the average monthly temperature of New York City during the years 2000, 2006, and 2012

Now we have three plots all on one graph. Python automatically chooses a different color for each line to indicate that the lines have been plotted from different data sets.

Instead of calling the plot function with all three pairs at once, we could also call the plot function three separate times, once for each pair:

```
>>> plot(months, nyc_temp_2000)
[<matplotlib.lines.Line2D object at 0x7f1e51351810>]
>>> plot(months, nyc_temp_2006)
[<matplotlib.lines.Line2D object at 0x7f1e5ae8e390>]
>>> plot(months, nyc_temp_2012)
[<matplotlib.lines.Line2D object at 0x7f1e5136ccd0>]
>>> show()
```

Matplotlib keeps track of what plots haven't been displayed yet. So as long as we wait to call show() until after we call plot() all three times, the plots will all get displayed on the same graph.

We have a problem, however, because we don't have any clue as to which color corresponds to which year. To fix this, we can use the function legend(), which lets us add a legend to the graph. A *legend* is a small display box that identifies what different parts of the graph mean. Here, we'll use a legend to indicate which year each colored line stands for. To add the legend, first call the plot() function as earlier:

```
>>> plot(months, nyc_temp_2000, months, nyc_temp_2006, months, nyc_temp_2012)
[<matplotlib.lines.Line2D object at 0x7f2549d6c410>, <matplotlib.lines.Line2D
object at 0x7f2549d6c9d0>, <matplotlib.lines.Line2D object at 0x7f2549a86850>]
```

Then, import the legend() function from the pylab module and call it as follows:

```
>>> from pylab import legend
>>> legend([2000, 2006, 2012])
<matplotlib.legend.Legend object at 0x7f2549d79410>
```

We call the legend() function with a list of the labels we want to use to identify each plot on the graph. These labels are entered in this order to match the order of the pairs of lists that were entered in the plot() function. That is, 2000 will be the label for the plot of the first pair we entered in the plot() function; 2006, for the second pair; and 2012, for the third. You can also specify a second argument to the function that will specify the position of the legend. By default, it's always positioned at the top right of the graph. However, you can specify a particular position, such as 'lower center', 'center left', and 'upper left'. Or you can set the position to 'best', and the legend will be positioned so as not to interfere with the graph.

Finally, we call show() to display the graph:

```
>>> show()
```

As you can see in the graph (see Figure 2-9), there's now a legend box in the top-right corner. It tells us which line represents the average monthly temperature for the year 2000, which line represents the year 2006, and which line represents the year 2012.

Looking at the graph, you can conclude two interesting facts: the highest temperature for all three years was in and around July (corresponding to 7 on the *x*-axis), and it has been increasing from 2000 with a more dramatic rise between 2000 and 2006. Having all three lines plotted together in one graph makes it a lot easier to see these kinds of relationships. It's certainly clearer than just looking at a few long lists of numbers or even looking at three lines plotted on three separate graphs.

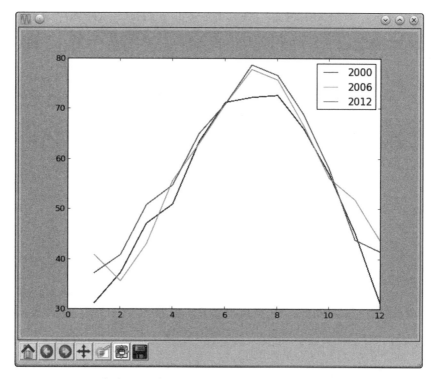

Figure 2-9: A graph showing the average monthly temperature of New York City, with a legend to show the year each color corresponds to

Customizing Graphs

We already learned about one way to customize a graph—by adding a legend. Now, we'll learn about other ways to customize a graph and to make it clearer by adding labels to the *x*- and *y*-axes, adding a title to the graph, and controlling the range and steps of the axes.

Adding a Title and Labels

We can add a title to our graph using the title() function and add labels for the *x*- and *y*-axes using the xlabel() and ylabel() functions. Let's re-create the last plot and add all this additional information:

```
>>> from pylab import plot, show, title, xlabel, ylabel, legend
>>> plot(months, nyc_temp_2000, months, nyc_temp_2006, months, nyc_temp_2012)
[<matplotlib.lines.Line2D object at 0x7f2549a9e210>, <matplotlib.lines.Line2D
object at 0x7f2549a4be90>, <matplotlib.lines.Line2D object at 0x7f2549a82090>]
>>> title('Average monthly temperature in NYC')
<matplotlib.text.Text object at 0x7f25499f7150>
>>> xlabel('Month')
<matplotlib.text.Text object at 0x7f2549d79210>
>>> ylabel('Temperature')
<matplotlib.text.Text object at 0x7f2549b8b2d0>
```

```
>>> legend([2000, 2006, 2012])
<matplotlib.legend.Legend object at 0x7f2549a82910>
```

All three functions—title(), xlabel(), and ylabel()—are called with the corresponding text that we want to appear on the graph entered as strings. Calling the show() function will display the graph with all this newly added information (see Figure 2-10).

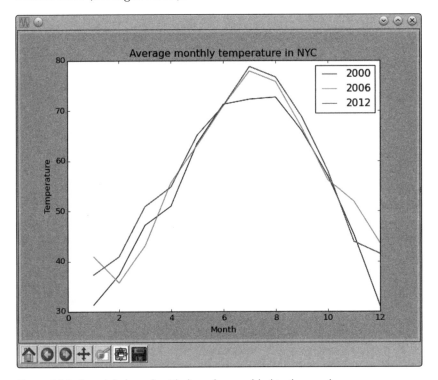

Figure 2-10: Axes labels and a title have been added to the graph.

With the three new pieces of information added, the graph is easier to understand.

Customizing the Axes

So far, we've allowed the numbers on both axes to be automatically determined by Python based on the data supplied to the plot() function. This may be fine for most cases, but sometimes this automatic range isn't the clearest way to present the data, as we saw in the graph where we plotted the average annual temperature of New York City (see Figure 2-7). There, even small changes in the temperature seemed large because the automatically chosen y-axis range was very narrow. We can adjust the range of the axes using the axis() function. This function can be used both to retrieve the current range and to set a new range for the axes.

Consider, once again, the average annual temperature of New York City during the years 2000 to 2012 and create a plot as we did earlier.

```
>>> nyc_temp = [53.9, 56.3, 56.4, 53.4, 54.5, 55.8, 56.8, 55.0, 55.3, 54.0, 56.7, 56.4, 57.3]
>>> plot(nyc_temp, marker='o')
[<matplotlib.lines.Line2D object at 0x7f3ae5b767d0>]
```

Now, import the axis() function and call it:

```
>>> from pylab import axis
>>> axis()
(0.0, 12.0, 53.0, 57.5)
```

The function returned a tuple with four numbers corresponding to the range for the *x*-axis (0.0, 12.0) and the *y*-axis (53.0, 57.5). These are the same range values from the graph that we made earlier. Now, let's change the *y*-axis to start from 0 instead of 53.0:

```
>>> axis(ymin=0)
(0.0, 12.0, 0, 57.5)
```

Calling the axis() function with the new starting value for the *y*-axis (specified by ymin=0) changes the range, and the returned tuple confirms it. If you display the graph by calling the show() function, the *y*-axis starts at 0, and the differences between the values of the consecutive years look less drastic (see Figure 2-11).

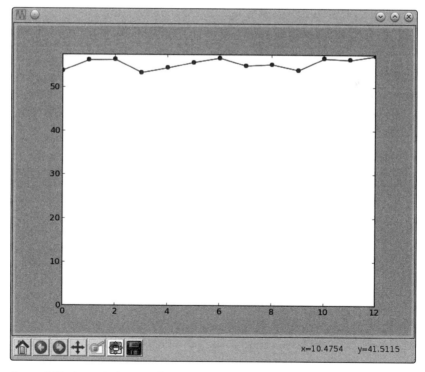

Figure 2-11: A graph showing the average annual temperature of New York City during the years 2000–2012. The y-axis has been customized to start from 0.

Similarly, you can use xmin, xmax, and ymax to set the minimum and maximum values for the *x*-axis and the maximum value for the *y*-axis, respectively. If you're changing all four values, you may find it easier to call the axis() function with all four range values entered as a list, such as axis([0, 10, 0, 20]). This would set the range of the *x*-axis to (0, 10) and that of the *y*-axis to (0, 20).

Plotting Using pyplot

The pylab module is useful for creating plots in an interactive shell, such as the IDLE shell, as we've been doing so far. However, when using matplotlib outside of the IDLE shell—for example, as part of a larger program—the pyplot module is more efficient. Don't worry—all the methods that you learned about when using pylab will work the same way with pyplot.

The following program recreates the first plot in this chapter using the pyplot module:

```
'''
Simple plot using pyplot
'''
❶ import matplotlib.pyplot

❷ def create_graph():
    x_numbers = [1, 2, 3]
    y_numbers = [2, 4, 6]

    matplotlib.pyplot.plot(x_numbers, y_numbers)
    matplotlib.pyplot.show()

if __name__ == '__main__':
    create_graph()
```

First, we import the pyplot module using the statement import matplotlib.pyplot ❶. This means that we're importing the entire pyplot module from the matplotlib package. To refer to any function or class definition defined in this module, you'll have to use the syntax matplotlib.pyplot.*item*, where *item* is the function or class you want to use.

This is different from importing a single function or class at a time, which is what we've been doing so far. For example, in the first chapter we imported the Fraction class as from fractions import Fraction. Importing an entire module is useful when you're going to use a number of functions from that module. Instead of importing them individually, you can just import the whole module at once and refer to different functions when you need them.

In the create_graph() function at ❷, we create the two lists of numbers that we want to plot on the graph and then pass the two lists to the plot() function, the same way we did before with pylab. This time, however, we call the function as matplotlib.pyplot.plot(), which means that we're calling the plot() function defined in the pyplot module of the matplotlib package. Then, we call the show() function to display the graph. The only difference between the way you plot the numbers here compared to what we did earlier is the mechanism of calling the functions.

To save us some typing, we can import the pyplot module by entering import matplotlib.pyplot as plt. Then, we can refer to pyplot with the label plt in our programs, instead of having to always type matplotlib.pyplot:

```
'''
Simple plot using pyplot
'''

import matplotlib.pyplot as plt

def create_graph():
    x_numbers = [1, 2, 3]
    y_numbers = [2, 4, 6]
    plt.plot(x_numbers, y_numbers)
    plt.show()

if __name__ == '__main__':
    create_graph()
```

Now, we can call the functions by prefixing them with the shortened plt instead of matplotlib.pyplot.

Going ahead, for the rest of this chapter and this book, we'll use pylab in the interactive shell and pyplot otherwise.

Saving the Plots

If you need to save your graphs, you can do so using the savefig() function. This function saves the graph as an image file, which you can use in reports or presentations. You can choose among several image formats, including PNG, PDF, and SVG.

Here's an example:

```
>>> from pylab import plot, savefig
>>> x = [1, 2, 3]
>>> y = [2, 4, 6]
>>> plot(x, y)
>>> savefig('mygraph.png')
```

This program will save the graph to an image file, *mygraph.png*, in your current directory. On Microsoft Windows, this is usually *C:\Python33* (where you installed Python). On Linux, the current directory is usually your home directory (*/home/<username>*), where *<username>* is the user you're logged in as. On a Mac, IDLE saves files to *~/Documents* by default. If you wanted to save it in a different directory, specify the complete pathname. For example, to save the image under *C:* on Windows as *mygraph.png*, you'd call the savefig() function as follows:

```
>>> savefig('C:\mygraph.png')
```

If you open the image in an image-viewing program, you'll see the same graph you'd see by calling the show() function. (You'll notice that the image file contains only the graph—not the entire window that pops up with the show() function). To specify a different image format, simply name the file with the appropriate extension. For example, mygraph.svg will create an SVG image file.

Another way to save a figure is to use the Save button in the window that pops up when you call show().

Plotting with Formulas

Until now, we've been plotting points on our graphs based on observed scientific measurements. In those graphs, we already had all our values for x and y laid out. For example, recorded temperatures and dates were already available to us at the time we wanted to create the New York City graph, showing how the temperature varied over months or years. In this section, we're going to create graphs from mathematical formulas.

Newton's Law of Universal Gravitation

According to Newton's law of universal gravitation, a body of mass m_1 attracts another body of mass m_2 with an amount of force F according to the formula

$$F = \frac{Gm_1m_2}{r^2},$$

where r is the distance between the two bodies and G is the gravitational constant. We want to see what happens to the force as the distance between the two bodies increases.

Let's take the masses of two bodies: the mass of the first body (m_1) is 0.5 kg, and the mass of the second body (m_2) is 1.5 kg. The value of the gravitational constant is 6.674×10^{-11} N m^2 kg^{-2}. Now we're ready to calculate the gravitational force between these two bodies at 19 different distances: 100 m, 150 m, 200 m, 250 m, 300 m, and so on up through 1000 m. The following program performs these calculations and also draws the graph:

```
'''
The relationship between gravitational force and
distance between two bodies
'''

import matplotlib.pyplot as plt

# Draw the graph
def draw_graph(x, y):
    plt.plot(x, y, marker='o')
    plt.xlabel('Distance in meters')
```

```
        plt.ylabel('Gravitational force in newtons')
        plt.title('Gravitational force and distance')
        plt.show()

    def generate_F_r():
        # Generate values for r
❶      r = range(100, 1001, 50)
        # Empty list to store the calculated values of F
        F = []

        # Constant, G
        G = 6.674*(10**-11)
        # Two masses
        m1 = 0.5
        m2 = 1.5

        # Calculate force and add it to the list, F
❷      for dist in r:
            force = G*(m1*m2)/(dist**2)
            F.append(force)

        # Call the draw_graph function
❸      draw_graph(r, F)

    if __name__=='__main__':
        generate_F_r()
```

The generate_F_r() function does most of the work in the program above. At ❶, we use the range() function to create a list labeled r with different values for distance, using a step value of 50. The final value is specified as 1001 because we want 1000 to be included as well. We then create an empty list (F), where we'll store the corresponding gravitational force at each of these distances. Next, we create labels referring to the gravitational constant (G) and the two masses (m1 and m2). Using a for loop ❷, we then calculate the force at each of the values in the list of distances (r). We use a label (force) to refer to the force calculated and to append it to the list (F). Finally, we call the function draw_graph() at ❸ with the list of distances and the list of the calculated forces. The x-axis of the graph displays the force, and the y-axis displays the distance. The graph is shown in Figure 2-12.

As the distance (r) increases, the gravitational force decreases. With this kind of relationship, we say that the gravitational force is *inversely proportional* to the distance between the two bodies. Also, note that when the value of one of the two variables changes, the other variable won't necessarily change by the same proportion. We refer to this as a *nonlinear relationship*. As a result, we end up with a curved line on the graph instead of a straight one.

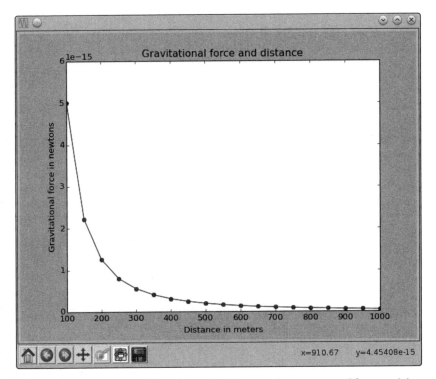

Figure 2-12: Visualization of the relationship between the gravitational force and the squared distance

Projectile Motion

Now, let's graph something you'll be familiar with from everyday life. If you throw a ball across a field, it follows a trajectory like the one shown in Figure 2-13.

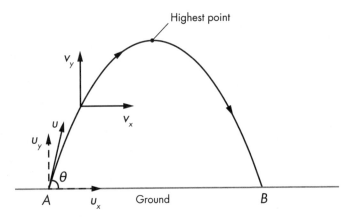

Figure 2-13: Motion of a ball that's thrown at point A—at an angle (θ) with a velocity (u)—and that hits the ground at point B

In the figure, the ball is thrown from point *A* and lands at point *B*. This type of motion is referred to as *projectile* motion. Our aim here is to use the equations of projectile motion to graph the trajectory of a body, showing the position of the ball starting from the point it's thrown until it hits the ground again.

When you throw the ball, it has an initial velocity and the direction of that velocity creates a certain angle with the ground. Let's call the initial velocity *u* and the angle that it makes with the ground θ (theta), as shown in Figure 2-13. The ball has two velocity components: one along the *x* direction, calculated by $u_x = u\cos\theta$, and the other along the *y* direction, where $u_y = u\sin\theta$.

As the ball moves, its velocity changes, and we will represent that changed velocity using *v*: the horizontal component is v_x and the vertical component is v_y. For simplicity, assume the horizontal component (v_x) doesn't change during the motion of the body, whereas the vertical component (v_y) decreases because of the force of gravity according to the equation $v_y = u_y - gt$. In this equation, *g* is the gravitational acceleration and *t* is the time at which the velocity is measured. Because $u_y = u\sin\theta$, we can substitute to get

$$v_y = u\sin\theta - gt.$$

Because the horizontal component of the velocity remains constant, the horizontal distance traveled (S_x) is given by $S_x = u(\cos\theta)t$. The vertical component of the velocity changes, though, and the vertical distance traveled is given by the formula

$$S_y = u(\sin\theta)t - \frac{1}{2}gt^2.$$

In other words, S_x and S_y give us the *x*- and *y*-coordinates of the ball at any given point in time during its flight. We'll use these equations when we write a program to draw the trajectory. As we use these equations, time (*t*) will be expressed in seconds, the velocity will be expressed in m/s, the angle of projection (θ) will be expressed in degrees, and the gravitational acceleration (*g*) will be expressed in m/s².

Before we write our program, however, we'll need to find out how long the ball will be in flight before it hits the ground so that we know when our program should stop plotting the trajectory of the ball. To do so, we'll first find how long the ball takes to reach its highest point. The ball reaches its highest point when the vertical component of the velocity (v_y) is 0, which is when $v_y = u\sin\theta - gt = 0$. So we're looking for the value *t* using the formula

$$t = \frac{u\sin\theta}{g}.$$

We'll call this time t_peak. After it reaches its highest point, the ball will hit the ground after being airborne for another t_peak seconds, so the total time of flight (t_flight) of the ball is

$$t_{\text{flight}} = 2t_{\text{peak}} = 2\frac{u\sin\theta}{g}.$$

Let's take a ball that's thrown with an initial velocity (u) of 5 m/s at an angle (θ) of 45 degrees. To calculate the total time of flight, we substitute $u = 5$, $\theta = 45$, and $g = 9.8$ into the equation we saw above:

$$t_{\text{flight}} = 2\frac{5\sin 45}{9.8}.$$

In this case, the time of flight for the ball turns out to be 0.72154 seconds (rounded to five decimal places). The ball will be in air for this period of time, so to draw the trajectory, we'll calculate its x- and y-coordinates at regular intervals during this time period. How often should we calculate the coordinates? Ideally, as frequently as possible. In this chapter, we'll calculate the coordinates every 0.001 seconds.

Generating Equally Spaced Floating Point Numbers

We've used the range() function to generate equally spaced integers— that is, if we wanted a list of integers between 1 and 10 with each integer separated by 1, we would use range(1, 10). If we wanted a different step value, we could specify that to the range function as the third argument. Unfortunately, there's no such built-in function for floating point numbers. So, for example, there's no function that would allow us to create a list of the numbers from 0 to 0.72 with two consecutive numbers separated by 0.001. We can use a while loop as follows to create our own function for this:

```
'''
Generate equally spaced floating point
numbers between two given values
'''

def frange(start, final, increment):

    numbers = []
❶    while start < final:
❷        numbers.append(start)
        start = start + increment

    return numbers
```

We've defined a function frange() ("floating point" range) that receives three parameters: start and final refer to the starting and the final points of the range of numbers, and increment refers to the difference between two consecutive numbers. We initialize a while loop at ❶, which continues execution as long as the number referred to by start is less than the value for final. We store the number pointed to by start in the list numbers ❷ and then add the value we entered as an increment during every iteration of the loop. Finally, we return the list numbers.

We'll use this function to generate equally spaced time instants in the trajectory-drawing program described next.

Drawing the Trajectory

The following program draws the trajectory of a ball thrown with a certain velocity and angle—both of which are supplied as input to the program:

```
'''
Draw the trajectory of a body in projectile motion
'''

from matplotlib import pyplot as plt
import math

def draw_graph(x, y):
    plt.plot(x, y)
    plt.xlabel('x-coordinate')
    plt.ylabel('y-coordinate')
    plt.title('Projectile motion of a ball')


def frange(start, final, interval):

    numbers = []
    while start < final:
        numbers.append(start)
        start = start + interval

    return numbers

def draw_trajectory(u, theta):

    theta = math.radians(theta)
    g = 9.8

    # Time of flight
    t_flight = 2*u*math.sin(theta)/g
    # Find time intervals
    intervals = frange(0, t_flight, 0.001)
```

❶ `theta = math.radians(theta)`

❷ `t_flight = 2*u*math.sin(theta)/g`

```
      # List of x and y coordinates
      x = []
      y = []
❸     for t in intervals:
          x.append(u*math.cos(theta)*t)
          y.append(u*math.sin(theta)*t - 0.5*g*t*t)

      draw_graph(x, y)

  if __name__ == '__main__':
❹     try:
          u = float(input('Enter the initial velocity (m/s): '))
          theta = float(input('Enter the angle of projection (degrees): '))
      except ValueError:
          print('You entered an invalid input')
      else:
          draw_trajectory(u, theta)
          plt.show()
```

In this program, we use the functions radians(), cos(), and sin() defined in the standard library's math module, so we import that module at the beginning. The draw_trajectory() function accepts two arguments, u and theta, corresponding to the velocity and the angle at which the ball is thrown. The math module's sine and the cosine functions expect the angle to be supplied in radians, so at ❶, we convert the angle (theta) from degrees to radians using the math.radians() function. Next, we create a label (g) to refer to the value of acceleration due to gravity, 9.8 m/s^2. At ❷, we calculate the time of flight and then call the frange() function with the values for start, final, and increment set to 0, t_flight, and 0.001, respectively. We then calculate the x- and y-coordinates for the trajectory at each of the time instants and store them in two separate lists, x and y ❸. To calculate these coordinates, we use the formulas for the distances S_x and S_y that we discussed earlier.

Finally, we call the draw_graph() function with the x- and y-coordinates to draw the trajectory. Note that the draw_graph() function doesn't call the show() function (we'll see why in the next program). We use a try...except block ❹ to report an error message in case the user enters an invalid input. Valid input for this program is any integer or floating point number. When you run the program, it asks for these values as input and then draws the trajectory (see Figure 2-14):

```
Enter the initial velocity (m/s): 25
Enter the angle of projection (degrees): 60
```

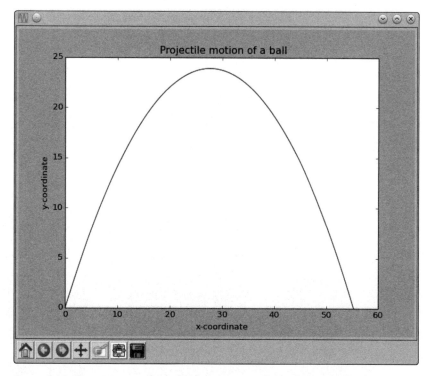

Figure 2-14: The trajectory of a ball when thrown with a velocity of 25 m/s at an angle of 60 degrees

Comparing the Trajectory at Different Initial Velocities

The previous program allows you to perform interesting experiments. For example, what will the trajectory look like for three balls thrown at different velocities but with the same initial angle? To graph three trajectories at once, we can replace the main code block from our previous program with the following:

```
if __name__ == '__main__':

      # List of three different initial velocities
❶    u_list = [20, 40, 60]
      theta = 45
      for u in u_list:
          draw_trajectory(u, theta)

      # Add a legend and show the graph
❷    plt.legend(['20', '40', '60'])
      plt.show()
```

Here, instead of asking the program's user to enter the velocity and the angle of projection, we create a list (u_list) with the velocities 20, 40, and 60 at ❶ and set the angle of projection as 45 degrees (using the label theta). We then call the draw_trajectory() function with each of the three values in u_list using the same value for theta, which calculates the list of *x*- and *y*-coordinates and calls the draw_graph() function. When we call the show() function, all three plots are displayed on the same graph. Because we now have a graph with multiple plots, we add a legend to the graph at ❷ before calling show() to display the velocity for each line. When you run the above program, you'll see the graph shown in Figure 2-15.

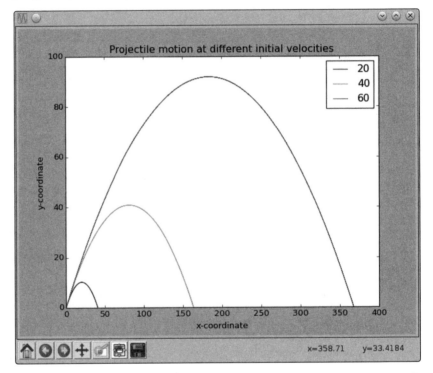

Figure 2-15: The trajectory of a ball thrown at a 60-degree angle, with a velocity of 20, 40, and 60 m/s

What You Learned

In this chapter, you learned the basics of creating graphs with matplotlib. You saw how to plot a single set of values, how to create multiple plots on the same graph, and how to label various parts of a graph to make it more informative. You used graphs to analyze the temperature variation of a city, study Newton's law of universal gravitation, and study the projectile motion of a body. In the next chapter, you'll use Python to start exploring statistics, and you'll see how drawing a graph can help make the relationships among sets of numbers easier to understand.

Programming Challenges

Here are a few challenges that build on what you've learned in this chapter. You can find sample solutions at *http://www.nostarch.com/doingmathwithpython/*.

#1: How Does the Temperature Vary During the Day?

If you enter a search term like "New York weather" in Google's search engine, you'll see, among other things, a graph showing the temperature at different times of the present day. Your task here is to re-create such a graph.

Using a city of your choice, find the temperature at different points of the day. Use the data to create two lists in your program and to create a graph with the time of day on the *x*-axis and the corresponding temperature on the *y*-axis. The graph should tell you how the temperature varies with the time of day. Try a different city and see how the two cities compare by plotting both lines on the same graph.

The time of day may be indicated by strings such as `'10:11 AM'` or `'09:21 PM'`.

#2: Exploring a Quadratic Function Visually

In Chapter 1, you learned how to find the roots of a quadratic equation, such as $x^2 + 2x + 1 = 0$. We can turn this equation into a function by writing it as $y = x^2 + 2x + 1$. For any value of *x*, the quadratic function produces *some* value for *y*. For example, when $x = 1$, $y = 4$. Here's a program that calculates the value of *y* for six different values of *x*:

```
'''
Quadratic function calculator
'''

# Assume values of x
❶ x_values = [-1, 1, 2, 3, 4, 5]
❷ for x in x_values:
    # Calculate the value of the quadratic function
    y = x**2 + 2*x + 1
    print('x={0} y={1}'.format(x, y))
```

At ❶, we create a list with six different values for x. The for loop starting at ❷ calculates the value of the function above for each of these values and uses the label y to refer to the list of results. Next, we print the value of x and the corresponding value of y. When you run the program, you should see the following output:

```
x=-1 y=0
x=1 y=4
x=2 y=9
```

```
x=3  y=16
x=4  y=25
x=5  y=36
```

Notice that the first line of the output is a root of the quadratic equation because it's a value for x that makes the function equal to 0.

Your programming challenge is to enhance this program to create a graph of the function. Try using at least 10 values for x instead of the 6 above. Calculate the corresponding y values using the function and then create a graph using these two sets of values.

Once you've created the graph, spend some time analyzing how the value of *y* varies with respect to *x*. Is the variation linear or nonlinear?

#3: Enhanced Projectile Trajectory Comparison Program

Your challenge here is to enhance the trajectory comparison program in a few ways. First, your program should print the time of flight, maximum horizontal distance, and maximum vertical distance traveled for each of the velocity and angle of projection combinations.

The other enhancement is to make the program work with any number of initial velocity and angle of projection values, supplied by the user. For example, here's how the program should ask the user for the inputs:

```
How many trajectories? 3
Enter the initial velocity for trajectory 1 (m/s): 45
Enter the angle of projection for trajectory 1 (degrees): 45
Enter the initial velocity for trajectory 2 (m/s): 60
Enter the angle of projection for trajectory 2 (degrees): 45
Enter the initial velocity for trajectory(m/s) 3: 45
Enter the angle of projection for trajectory(degrees) 3: 90
```

Your program should also ensure that erroneous input is properly handled using a try...except block, just as in the original program.

#4: Visualizing Your Expenses

I always find myself asking at the end of the month, "Where did all that money go?" I'm sure this isn't a problem I alone face.

For this challenge, you'll write a program that creates a bar chart for easy comparison of weekly expenditures. The program should first ask for the number of categories for the expenditures and the weekly total expenditure in each category, and then it should create the bar chart showing these expenditures.

Here's a sample run of how the program should work:

```
Enter the number of categories: 4
Enter category: Food
Expenditure: 70
```

```
Enter category: Transportation
Expenditure: 35
Enter category: Entertainment
Expenditure: 30
Enter category: Phone/Internet
Expenditure: 30
```

Figure 2-16 shows the bar chart that will be created to compare the expenditures. If you save the bar chart for every week, at the end of the month, you'll be able to see how the expenditures varied between the weeks for different categories.

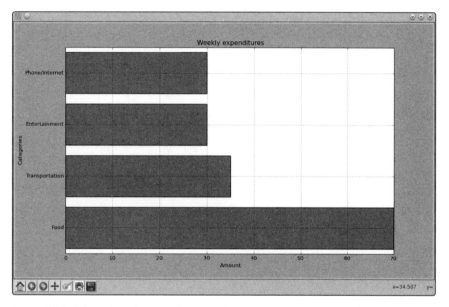

Figure 2-16: A bar chart showing the expenditures per category during the week

We haven't discussed creating a bar chart using matplotlib, so let's try an example.

A bar chart can be created using matplotlib's barh() function, which is also defined in the pyplot module. Figure 2-17 shows a bar chart that illustrates the number of steps I walked during the past week. The days of the week—Sunday, Monday, Tuesday, and so forth—are referred to as the *labels*. Each horizontal bar starts from the *y*-axis, and we have to specify the *y*-coordinate of the *center* of this position for each of the bars. The length of each bar corresponds to the number of steps specified.

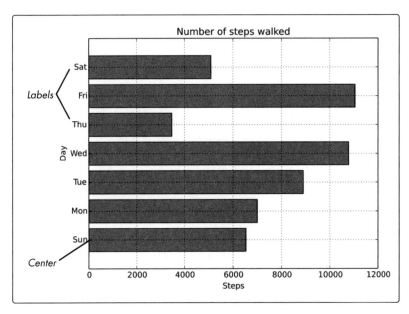

Figure 2-17: A bar chart showing the number of steps walked during a week

The following program creates the bar chart:

```
'''
Example of drawing a horizontal bar chart
'''
import matplotlib.pyplot as plt
def create_bar_chart(data, labels):
    # Number of bars
    num_bars = len(data)
    # This list is the point on the y-axis where each
    # Bar is centered. Here it will be [1, 2, 3...]
    positions = range(1, num_bars+1)
    plt.barh(positions, data, align='center')
    # Set the label of each bar
    plt.yticks(positions, labels)
    plt.xlabel('Steps')
    plt.ylabel('Day')
    plt.title('Number of steps walked')
    # Turns on the grid which may assist in visual estimation
    plt.grid(
    plt.show()

if __name__ == '__main__':
    # Number of steps I walked during the past week
    steps = [6534, 7000, 8900, 10786, 3467, 11045, 5095]
    # Corresponding days
    labels = ['Sun', 'Mon', 'Tue', 'Wed', 'Thu', 'Fri', 'Sat']
    create_bar_chart(steps, labels)
```

❶
❷

The create_bar_chart() function accepts two parameters—data, which is a list of numbers we want to represent using the bars and labels, and the corresponding labels list. The center of each bar has to be specified, and I've arbitrarily chosen the centers as 1, 2, 3, 4, and so on using the help of the range() function at ❶.

We then call the barh() function, passing positions and data as the first two arguments and then the keyword argument, align='center', at ❷. The keyword argument specifies that the bars are centered at the positions on the y-axis specified by the list. We then set the labels for each bar, the axis labels, and the title using the yticks() function. We also call the grid() function to turn on the grid, which may be useful for a visual estimation of the number of steps. Finally, we call the show() function.

#5: Exploring the Relationship Between the Fibonacci Sequence and the Golden Ratio

The Fibonacci sequence (1, 1, 2, 3, 5, . . .) is the series of numbers where the ith number in the series is the sum of the two previous numbers—that is, the numbers in the positions $(i - 2)$ and $(i - 1)$. The successive numbers in this series display an interesting relationship. As you increase the number of terms in the series, the ratios of consecutive pairs of numbers are nearly equal to each other. This value approaches a special number referred to as the *golden ratio*. Numerically, the golden ratio is the number 1.618033988 . . . , and it's been the subject of extensive study in music, architecture, and nature. For this challenge, write a program that will plot on a graph the ratio between consecutive Fibonacci numbers for, say, 100 numbers, which will demonstrate that the values approach the golden ratio.

You may find the following function, which returns a list of the first n Fibonacci numbers, useful in implementing your solution:

```
def fibo(n):
    if n == 1:
        return [1]
    if n == 2:
        return [1, 1]
    # n > 2
    a = 1
    b = 1
    # First two members of the series
    series = [a, b]
    for i in range(n):
        c = a + b
        series.append(c)
        a = b
        b = c

    return series
```

The output of your solution should be a graph, as shown in Figure 2-18.

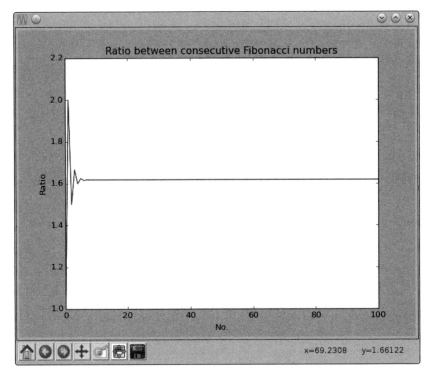

Figure 2-18: The ratio between the consecutive Fibonacci numbers approaches the golden ratio.

3

DESCRIBING DATA WITH STATISTICS

In this chapter, we'll use Python to explore statistics so we can study, describe, and better understand sets of data. After looking at some basic statistical measures—the mean, median, mode, and range—we'll move on to some more advanced measures, such as variance and standard deviation. Then, we'll see how to calculate the correlation coefficient, which allows you to quantify the relationship between two sets of data. We'll end the chapter by learning about scatter plots. Along the way, we'll learn more about the Python language and standard library modules. Let's get started with one of the most commonly used statistical measures—the mean.

NOTE   *In statistics, some statistical measures are calculated slightly differently depending on whether you have data for an entire population or just a sample. To keep things simple, we'll stick with the calculation methods for a population in this chapter.*

Finding the Mean

The *mean* is a common and intuitive way to summarize a set of numbers. It's what we might simply call the "average" in everyday use, although as we'll see, there are other kinds of averages as well. Let's take a sample set of numbers and calculate the mean.

Say there's a school charity that's been taking donations over a period of time spanning the last 12 days (we'll refer to this as period A). In that time, the following 12 numbers represent the total dollar amount of donations received for each day: 100, 60, 70, 900, 100, 200, 500, 500, 503, 600, 1000, and 1200. We can calculate the mean by summing these totals and then dividing the sum by the number of days. In this case, the sum of the numbers is 5733. If we divide this number by 12 (the number of days), we get 477.75, which is the *mean* donation per day. This number gives us a general idea of how much money was donated on any given day.

In a moment, we'll write a program that calculates and prints the mean for a collection of numbers. As we just saw, to calculate the mean, we'll need to take the sum of the list of numbers and divide it by the number of items in the list. Let's look at two Python functions that make both of these operations very easy: sum() and len().

When you use the sum() function on a list of numbers, it adds up all the numbers in the list and returns the result:

```
>>> shortlist = [1, 2, 3]
>>> sum(shortlist)
6
```

We can use the len() function to give us the length of a list:

```
>>> len(shortlist)
3
```

When we use the len() function on the list, it returns 3 because there are three items in shortlist. Now we're ready to write a program that will calculate the mean of the list of donations.

```
'''
Calculating the mean
'''

def calculate_mean(numbers):
❶    s = sum(numbers)
❷    N = len(numbers)
    # Calculate the mean
❸    mean = s/N

    return mean
```

```
    if __name__ == '__main__':
❹      donations = [100, 60, 70, 900, 100, 200, 500, 500, 503, 600, 1000, 1200]
❺      mean = calculate_mean(donations)
       N = len(donations)
❻      print('Mean donation over the last {0} days is {1}'.format(N, mean))
```

First, we define a function, calculate_mean(), that accepts the argument numbers, which is a list of numbers. At ❶, we use the sum() function to add up the numbers in the list and create a label, s, to refer to the total. Similarly, at ❷, we use the len() function to get the length of the list and create a label, N, to refer to it. Then, as you can see at ❸, we calculate the mean by simply dividing the sum (s) by the number of members (N). At ❹, we create a list, donations, with the values of the donations listed earlier. We then call the calculate_mean() function, passing this list as an argument at ❺. Finally, we print the mean that was calculated at ❻.

When you run the program, you should see the following:

```
Mean donation over the last 12 days is 477.75
```

The calculate_mean() function will calculate the sum and length of *any* list, so we can reuse it to calculate the mean for other sets of numbers, too.

We calculated that the mean donation per day was 477.75. It's worth noting that the donations during the first few days were much lower than the mean donation we calculated and that the donations during the last couple of days were much higher. The mean gives us one way to summarize the data, but it doesn't give us a full picture. There are other statistical measurements, however, that can tell us more about the data when compared with the mean.

Finding the Median

The *median* of a collection of numbers is another kind of average. To find the median, we sort the numbers in ascending order. If the length of the list of numbers is odd, the number in the middle of the list is the median. If the length of the list of numbers is even, we get the median by taking the mean of the two middle numbers. Let's find the median of the previous list of donations: 100, 60, 70, 900, 100, 200, 500, 500, 503, 600, 1000, and 1200.

After sorting from smallest to largest, the list of numbers becomes 60, 70, 100, 100, 200, 500, 500, 503, 600, 900, 1000, and 1200. We have an even number of items in the list (12), so to get the median, we need to take the mean of the two middle numbers. In this case, the middle numbers are the sixth and the seventh numbers—500 and 500—and the mean of these two numbers is (500 + 500)/2, which comes out to 500. That means the median is 500.

Now assume—just for this example—that we have another donation total for the 13th day so that the list now looks like this: 100, 60, 70, 900, 100, 200, 500, 500, 503, 600, 1000, 1200, and 800.

Once again, we have to sort the list, which becomes 60, 70, 100, 100, 200, 500, 500, 503, 600, 800, 900, 1000, and 1200. There are 13 numbers in this list (an odd number), so the median for this list is simply the middle number. In this case, that's the seventh number, which is 500.

Before we write a program to find the median of a list of numbers, let's think about how we could automatically calculate the middle elements of a list in either case. If the length of a list (N) is odd, the middle number is the one in position $(N+1)/2$. If N is even, the two middle elements are $N/2$ and $(N/2) + 1$. For our first example in this section, $N = 12$, so the two middle elements were the $12/2$ (sixth) and $12/2 + 1$ (seventh) elements. In the second example, $N = 13$, so the seventh element, $(N+1)/2$, was the middle element.

In order to write a function that calculates the median, we'll also need to sort a list in ascending order. Luckily, the sort() method does just that:

```
>>> samplelist = [4, 1, 3]
>>> samplelist.sort()
>>> samplelist
[1, 3, 4]
```

Now we can write our next program, which finds the median of a list of numbers:

```
'''
Calculating the median
'''

def calculate_median(numbers):
❶    N = len(numbers)
❷    numbers.sort()

    # Find the median
    if N % 2 == 0:
        # if N is even
        m1 = N/2
        m2 = (N/2) + 1
        # Convert to integer, match position
❸        m1 = int(m1) - 1
❹        m2 = int(m2) - 1
❺        median = (numbers[m1] + numbers[m2])/2
    else:
❻        m = (N+1)/2
        # Convert to integer, match position
        m = int(m) - 1
        median = numbers[m]

    return median

if __name__ == '__main__':
    donations = [100, 60, 70, 900, 100, 200, 500, 500, 503, 600, 1000, 1200]
```

```
median = calculate_median(donations)
N = len(donations)
print('Median donation over the last {0} days is {1}'.format(N, median))
```

The overall structure of the program is similar to that of the earlier program that calculates the mean. The calculate_median() function accepts a list of numbers and returns the median. At ❶, we calculate the length of the list and create a label, N, to refer to it. Next, at ❷, we sort the list using the sort() method.

Then, we check to see whether N is even. If so, we find the middle elements, m1 and m2, which are the numbers at positions N/2 and (N/2) + 1 in the sorted list. The next two statements (❸ and ❹) adjust m1 and m2 in two ways. First, we use the int() function to convert m1 and m2 into integer form. This is because results of the division operator are always returned as floating point numbers, even when the result is equivalent to an integer. For example:

```
>>> 6/2
3.0
```

We cannot use a floating point number as an index in a list, so we use int() to convert that result to an integer. We also subtract 1 from both m1 and m2 because positions in a list begin with 0 in Python. This means that to get the sixth and seventh numbers from the list, we have to ask for the numbers at index 5 and index 6. At ❺, we calculate the median by taking the mean of the two numbers in the middle positions.

Starting at ❻, the program finds the median if there's an odd number of items in the list, once again using int() and subtracting 1 to find the proper index. Finally, the program calculates the median for the list of donations and returns it. When you execute the program, it calculates that the median is 500:

```
Median donation over the last 12 days is 500.0
```

As you can see, the mean (477.75) and the median (500) are pretty close in this particular list, but the median is a little higher.

Finding the Mode and Creating a Frequency Table

Instead of finding the mean value or the median value of a set of numbers, what if you wanted to find the number that occurs most frequently? This number is called the *mode*. For example, consider the test scores of a math test (out of 10 points) in a class of 20 students: 7, 8, 9, 2, 10, 9, 9, 9, 9, 4, 5, 6, 1, 5, 6, 7, 8, 6, 1, and 10. The mode of this list would tell you which score was the most common in the class. From the list, you can see that the score of 9 occurs most frequently, so 9 is the mode for this list of numbers. There's no symbolic formula for calculating the mode—you simply count how many times each unique number occurs and find the one that occurs the most.

To write a program to calculate the mode, we'll need to have Python count how many times each number occurs within a list and print the one that occurs most frequently. The Counter class from the collections module, which is part of the standard library, makes this really simple for us.

Finding the Most Common Elements

Finding the most common number in a data set can be thought of as a subproblem of finding an arbitrary number of most common numbers. For instance, instead of the most common score, what if you wanted to know the five most common scores? The most_common() method of the Counter class allows us to answer such questions easily. Let's see an example:

```
>>> simplelist = [4, 2, 1, 3, 4]
>>> from collections import Counter
>>> c = Counter(simplelist)
>>> c.most_common()
[(4, 2), (1, 1), (2, 1), (3, 1)]
```

Here, we start off with a list of five numbers and import Counter from the collections module. Then, we create a Counter object, using c to refer to the object. We then call the most_common() method, which returns a list ordered by the most common elements.

Each member of the list is a tuple. The first element of the first tuple is the number that occurs most frequently, and the second element is the number of times it occurs. The second, third, and fourth tuples contain the other numbers along with the count of the number of times they appear. This result tells us that 4 occurs the most (twice), while the others appear only once. Note that numbers that occur an equal number of times are returned by the most_common() method in an arbitrary order.

When you call the most_common() method, you can also provide an argument telling it the number of most common elements you want it to return. For example, if we just wanted to find the most common element, we would call it with the argument 1:

```
>>> c.most_common(1)
[(4, 2)]
```

If you call the method again with 2 as an argument, you'll see this:

```
>>> c.most_common(2)
[(4, 2), (1, 1)]
```

Now the result returned by the most_common method is a list with two tuples. The first is the most common element, followed by the second most common. Of course, in this case, there are several elements tied for most common, so the fact that the function returns 1 here (and not 2 or 3) is arbitrary, as noted earlier.

The most_common() method returns both the numbers and the number of times they occur. What if we want only the numbers and we don't care about the number of times they occur? Here's how we can retrieve that information:

```
❶ >>> mode = c.most_common(1)
   >>> mode
   [(4, 2)]
❷ >>> mode[0]
   (4, 2)
❸ >>> mode[0][0]
   4
```

At ❶, we use the label mode to refer to the result returned by the most_common() method. We retrieve the first (and the only) element of this list with mode[0] ❷, which gives us a tuple. Because we just want the first element of the tuple, we can retrieve that using mode[0][0] ❸. This returns 4—the most common element, or the mode.

Now that we know how the most_common() method works, we'll apply it to solve the next two problems.

Finding the Mode

We're ready to write a program that finds the mode for a list of numbers:

```
   '''
   Calculating the mode
   '''

   from collections import Counter

   def calculate_mode(numbers):
❶      c = Counter(numbers)
❷      mode = c.most_common(1)
❸      return mode[0][0]

   if __name__=='__main__':
       scores = [7, 8, 9, 2, 10, 9, 9, 9, 9, 4, 5, 6, 1, 5, 6, 7, 8, 6, 1, 10]
       mode = calculate_mode(scores)

       print('The mode of the list of numbers is: {0}'.format(mode))
```

The calculate_mode() function finds and returns the mode of the numbers passed to it as a parameter. To calculate the mode, we first import the class Counter from the collections module and use it to create a Counter object at ❶. Then, at ❷, we use the most_common() method, which, as we saw earlier, gives us a list that contains a tuple with the most common number and the number of times it occurs. We assign that list the label mode. Finally, we use mode[0][0] ❸ to access the number we want: the most frequent number from the list, which is the mode.

The rest of the program applies the `calculate_mode` function to the list of test scores we saw earlier. When you run the program, you should see the following output:

```
The mode of the list of numbers is: 9
```

What if you have a set of data where two or more numbers occur the same maximum number of times? For example, in the list of numbers 5, 5, 5, 4, 4, 4, 9, 1, and 3, both 4 and 5 are present three times. In such cases, the list of numbers is said to have multiple modes, and our program should find and print all the modes. The modified program follows:

```
'''
Calculating the mode when the list of numbers may
have multiple modes
'''

from collections import Counter


def calculate_mode(numbers):

    c = Counter(numbers)
❶    numbers_freq = c.most_common()
❷    max_count = numbers_freq[0][1]

    modes = []
    for num in numbers_freq:
❸        if num[1] == max_count:
            modes.append(num[0])
    return modes

if __name__ == '__main__':
    scores = [5, 5, 5, 4, 4, 4, 9, 1, 3]
    modes = calculate_mode(scores)
    print('The mode(s) of the list of numbers are:')
❹    for mode in modes:
        print(mode)
```

At ❶, instead of finding only the most common element, we retrieve all the numbers and the number of times each appears. Next, at ❷, we find the value of the maximum count—that is, the maximum number of times any number occurs. Then, for each of the numbers, we check whether the number of times it appears is equal to the maximum count ❸. Each number that fulfills this condition is a mode, and we add it to the list modes and return the list.

At ❹, we iterate over the list returned from the `calculate_mode()` function and print each of the numbers.

When you execute the preceding program, you should see the following output:

```
The mode(s) of the list of numbers are:
4
5
```

What if you wanted to find the number of times every number occurs instead of just the mode? A *frequency table*, as the name indicates, is a table that shows how many times each number occurs within a collection of numbers.

Creating a Frequency Table

Let's consider the list of test scores again: 7, 8, 9, 2, 10, 9, 9, 9, 9, 4, 5, 6, 1, 5, 6, 7, 8, 6, 1, and 10. The frequency table for this list is shown in Table 3-1. For each number, we list the number of times it occurs in the second column.

Table 3-1: Frequency Table

Score	Frequency
1	2
2	1
4	1
5	2
6	3
7	2
8	2
9	5
10	2

Note that the sum of the individual frequencies in the second column adds up to the total number of scores (in this case, 20).

We'll use the most_common() method once again to print the frequency table for a given set of numbers. Recall that when we don't supply an argument to the most_common() method, it returns a list of tuples with all the numbers and the number of times they appear. We can simply print each number and its frequency from this list to display a frequency table.

Here's the program:

```
'''
Frequency table for a list of numbers
'''

from collections import Counter
```

```
def frequency_table(numbers):
❶    table = Counter(numbers)
     print('Number\tFrequency')
❷    for number in table.most_common():
         print('{0}\t{1}'.format(number[0], number[1]))

if __name__=='__main__':
    scores = [7, 8, 9, 2, 10, 9, 9, 9, 9, 4, 5, 6, 1, 5, 6, 7, 8, 6, 1, 10]
    frequency_table(scores)
```

The function frequency_table() prints the frequency table of the list of numbers passed to it. At ❶, we first create a Counter object and create the label table to refer to it. Next, using a for loop ❷, we go through each of the tuples, printing the first member (the number itself) and the second member (the frequency of the corresponding number). We use \t to print a tab between each value to space the table. When you run the program, you'll see the following output:

```
Number  Frequency
9       5
6       3
1       2
5       2
7       2
8       2
10      2
2       1
4       1
```

Here, you can see that the numbers are listed in decreasing order of frequency because the most_common() function returns the numbers in this order. If, instead, you want your program to print the frequency table sorted by value from lowest to highest, as shown in Table 3-1, you'll have to re-sort the list of tuples.

The sort() method is all we need to modify our earlier frequency table program:

```
'''
Frequency table for a list of numbers
Enhanced to display the table sorted by the numbers
'''

from collections import Counter

def frequency_table(numbers):
    table = Counter(numbers)
❶    numbers_freq = table.most_common()
❷    numbers_freq.sort()

    print('Number\tFrequency')
❸    for number in numbers_freq:
        print('{0}\t{1}'.format(number[0], number[1]))
```

```
if __name__ == '__main__':
    scores = [7, 8, 9, 2, 10, 9, 9, 9, 9, 4, 5, 6, 1, 5, 6, 7, 8, 6, 1, 10]
    frequency_table(scores)
```

Here, we store the list returned by the most_common() method in numbers_freq at ❶, and then we sort it by calling the sort() method ❷. Finally, we use the for loop to go over the sorted tuples and print each number and its frequency ❸. Now when you run the program, you'll see the following table, which is identical to Table 3-1:

Number	Frequency
1	2
2	1
4	1
5	2
6	3
7	2
8	2
9	5
10	2

In this section, we've covered mean, median, and mode, which are three common measures for describing a list of numbers. Each of these can be useful, but they can also hide other aspects of the data when considered in isolation. Next, we'll look at other, more advanced statistical measures that can help us draw more conclusions about a collection of numbers.

Measuring the Dispersion

The next statistical calculations we'll look at measure the *dispersion*, which tells us how far away the numbers in a set of data are from the mean of the data set. We'll learn to calculate three different measurements of dispersion: range, variance, and standard deviation.

Finding the Range of a Set of Numbers

Once again, consider the list of donations during period A: 100, 60, 70, 900, 100, 200, 500, 500, 503, 600, 1000, and 1200. We found that the mean donation per day is 477.75. But just looking at the mean, we have no idea whether all the donations fell into a narrow range—say between 400 and 500—or whether they varied much more than that—say between 60 and 1200, as in this case. For a list of numbers, the *range* is the difference between the highest number and the lowest number. You could have two groups of numbers with the exact same mean but with vastly different ranges, so knowing the range fills in more information about a set of numbers beyond what we can learn from just looking at the mean, median, and mode.

The next program finds the range of the preceding list of donations:

```
'''
Find the range
'''

def find_range(numbers):
❶    lowest = min(numbers)
❷    highest = max(numbers)
     # Find the range
     r = highest-lowest

❸    return lowest, highest, r


if __name__ == '__main__':
    donations = [100, 60, 70, 900, 100, 200, 500, 500, 503, 600, 1000, 1200]
❹   lowest, highest, r = find_range(donations)
    print('Lowest: {0} Highest: {1} Range: {2}'.format(lowest, highest, r))
```

The function find_range() accepts a list as a parameter and finds the range. First, it calculates the lowest and the highest numbers using the min() and the max() functions at ❶ and ❷. As the function names indicate, they find the minimum and the maximum values in a list of numbers.

We then calculate the range by taking the difference between the highest and the lowest numbers, using the label r to refer to this difference. At ❸, we return all three numbers—the lowest number, the highest number, and the range. This is the first time in the book that we're returning multiple values from a function—instead of just returning one value, this function returns three. At ❹, we use three labels to *receive* the three values being returned from the find_range() function. Finally, we print the values. When you run the program, you should see the following output:

```
Lowest: 60 Highest: 1200 Range: 1140
```

This tells us that the days' total donations were fairly spread out, with a range of 1140, because we had daily totals as small as 60 and as large as 1200.

Finding the Variance and Standard Deviation

The range tells us the difference between the two extremes in a set of numbers, but what if we want to know more about how all of the individual numbers vary from the mean? Were they all similar, clustered near the mean, or were they all different, closer to the extremes? There are two related measures of dispersion that tell us more about a list of numbers along these lines: the *variance* and the *standard deviation*. To calculate either of these, we first need to find the difference of each of the numbers from the mean. The variance is the average of the squares of those differences.

A high variance means that values are far from the mean; a low variance means that the values are clustered close to the mean. We calculate the variance using the formula

$$\text{variance} = \frac{\sum (x_i - x_{\text{mean}})^2}{n}.$$

In the formula, x_i stands for individual numbers (in this case, daily total donations), x_{mean} stands for the mean of these numbers (the mean daily donation), and n is the number of values in the list (the number of days on which donations were received). For each value in the list, we take the difference between that number and the mean and square it. Then, we add all those squared differences together and, finally, divide the whole sum by n to find the variance.

If we want to calculate the standard deviation as well, all we have to do is take the square root of the variance. Values that are within one standard deviation of the mean can be thought of as fairly typical, whereas values that are three or more standard deviations away from the mean can be considered much more atypical—we call such values *outliers*.

Why do we have these two measures of dispersion—variance and standard deviation? In short, the two measures are useful in different situations. Going back to the formula we used to calculate the variance, you can see that the variance is expressed in square units because it's the average of the squared difference from the mean. For some mathematical formulas, it's nicer to work with those square units instead of taking the square root to find the standard deviation. On the other hand, the standard deviation is expressed in the same units as the population data. For example, if you calculate the variance for our list of donations (as we will in a moment), the result is expressed in dollars squared, which doesn't make a lot of sense. Meanwhile, the standard deviation is simply expressed in dollars, the same unit as each of the donations.

The following program finds the variance and standard deviation for a list of numbers:

```
'''
Find the variance and standard deviation of a list of numbers
'''

def calculate_mean(numbers):
    s = sum(numbers)
    N = len(numbers)
    # Calculate the mean
    mean = s/N

    return mean

def find_differences(numbers):
    # Find the mean
    mean = calculate_mean(numbers)
    # Find the differences from the mean
    diff = []
```

```
    for num in numbers:
        diff.append(num-mean)

    return diff

def calculate_variance(numbers):

    # Find the list of differences
❶  diff = find_differences(numbers)
    # Find the squared differences
    squared_diff = []
❷  for d in diff:
        squared_diff.append(d**2)
    # Find the variance
    sum_squared_diff = sum(squared_diff)
❸  variance = sum_squared_diff/len(numbers)
    return variance

if __name__ == '__main__':
    donations = [100, 60, 70, 900, 100, 200, 500, 500, 503, 600, 1000, 1200]
    variance = calculate_variance(donations)
    print('The variance of the list of numbers is {0}'.format(variance))

❹  std = variance**0.5
    print('The standard deviation of the list of numbers is {0}'.format(std))
```

The function calculate_variance() calculates the variance of the list of numbers passed to it. First, it calls the find_differences() function at ❶ to calculate the difference of each of the numbers from the mean. The find_differences() function returns the difference of each donation from the mean value as a list. In this function, we use the calculate_mean() function we wrote earlier to find the mean donation. Then, starting at ❷, the squares of these differences are calculated and saved in a list labeled squared_diff. Next, we use the sum() function to find the sum of the squared differences and, finally, calculate the variance at ❸. At ❹, we calculate the standard deviation by taking the square root of the variance.

When you run the preceding program, you should see the following output:

```
The variance of the list of numbers is 141047.35416666666
The standard deviation of the list of numbers is 375.5627166887931
```

The variance and the standard deviation are both very large, meaning that the individual daily total donations vary greatly from the mean. Now, let's compare the variance and the standard deviation for a different set of donations that have the same mean: 382, 389, 377, 397, 396, 368, 369, 392, 398, 367, 393, and 396. In this case, the variance and the standard deviation turn out to be 135.38888888888889 and 11.63567311713804, respectively. Lower values for variance and standard deviation tell us that the individual numbers are closer to the mean. Figure 3-1 illustrates this point visually.

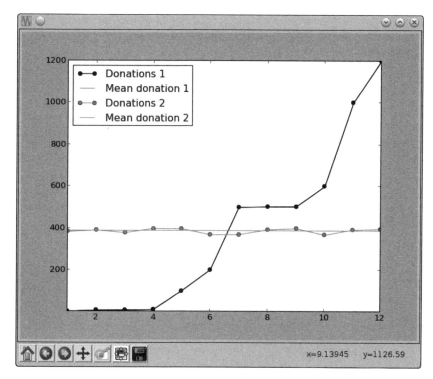

Figure 3-1: Variation of the donations around the average donation

The mean donations for both lists of donations are similar, so the two lines overlap, appearing as a single line in the figure. However, the donations from the first list vary widely from the mean, whereas the donations from the second list are very close to the mean, which confirms what we inferred from the lower variance value.

Calculating the Correlation Between Two Data Sets

In this section, we'll learn how to calculate a statistical measure that tells us the nature and strength of the relationship between two sets of numbers: the *Pearson correlation coefficient*, which I'll call simply the *correlation coefficient*. Note that this coefficient measures the strength of the *linear* relationship. We'd have to use other measures (which we won't be discussing here) to find out the coefficient when two sets have a nonlinear relationship. The coefficient can be either positive or negative, and its magnitude can range between −1 and 1 (inclusive).

A correlation coefficient of 0 indicates that there's no linear correlation between the two quantities. (Note that this doesn't mean the two quantities are independent of each other. There could still be a nonlinear relationship between them, for example). A coefficient of 1 or close to 1 indicates that there's a strong positive linear correlation; a coefficient of exactly 1 is referred to as perfect positive correlation. Similarly, a correlation coefficient of –1 or close to –1 indicates a strong negative correlation, where 1 indicates a perfect negative correlation.

CORRELATION AND CAUSATION

In statistics, you'll often come across the statement "correlation doesn't imply causation." This is a reminder that even if two sets of observations are strongly correlated with each other, that doesn't mean one variable *causes* the other. When two variables are strongly correlated, sometimes there's a third factor that influences both variables and explains the correlation. A classic example is the correlation between ice cream sales and crime rates—if you track both of these variables in a typical city, you're likely to find a correlation, but this doesn't mean that ice cream sales cause crime (or vice versa). Ice cream sales and crime are correlated because they both go up as the weather gets hotter during the summer. Of course, this doesn't mean that hot weather directly causes crime to go up either; there are more complicated causes behind that correlation as well.

Calculating the Correlation Coefficient

The correlation coefficient is calculated using the formula

$$\text{correlation} = \frac{n\sum xy - \sum x \sum y}{\sqrt{\left(n\sum x^2 - \left(\sum x\right)^2\right)\left(n\sum y^2 - \left(\sum y\right)^2\right)}}.$$

In the above formula, n is the total number of values present in each set of numbers (the sets have to be of equal length). The two sets of numbers are denoted by x and y (it doesn't matter which one you denote as which). The other terms are described as follows:

$\sum xy$	Sum of the products of the individual elements of the two sets of numbers, x and y
$\sum x$	Sum of the numbers in set x
$\sum y$	Sum of the numbers in set y
$\left(\sum x\right)^2$	Square of the sum of the numbers in set x
$\left(\sum y\right)^2$	Square of the sum of the numbers in set y
$\sum x^2$	Sum of the squares of the numbers in set x
$\sum y^2$	Sum of the squares of the numbers in set y

Once we've calculated these terms, you can combine them according to the preceding formula to find the correlation coefficient. For small lists, it's possible to do this by hand without too much effort, but it certainly gets complicated as the size of each set of numbers increases.

In a moment, we'll write a program that calculates the correlation coefficient for us. In this program, we'll use the zip() function, which will help us calculate the sum of products from the two sets of numbers. Here's an example of how the zip() function works:

```
>>> simple_list1 = [1, 2, 3]
>>> simple_list2 = [4, 5, 6]
>>> for x, y in zip(simple_list1, simple_list2):
        print(x, y)

1 4
2 5
3 6
```

The zip() function returns pairs of the corresponding elements in x and y, which you can then use in a loop to perform other operations (like printing, as shown in the preceding code). If the two lists are unequal in length, the function terminates when all the elements of the smaller list have been read.

Now we're ready to write a program that will calculate the correlation coefficient for us:

```
def find_corr_x_y(x,y):
    n = len(x)

    # Find the sum of the products
    prod = []
❶    for xi,yi in zip(x,y):
        prod.append(xi*yi)

❷    sum_prod_x_y = sum(prod)
❸    sum_x = sum(x)
❹    sum_y = sum(y)
    squared_sum_x = sum_x**2
    squared_sum_y = sum_y**2

    x_square = []
❺    for xi in x:
        x_square.append(xi**2)
    # Find the sum
    x_square_sum = sum(x_square)

    y_square=[]
     for yi in y:
        y_square.append(yi**2)
    # Find the sum
    y_square_sum = sum(y_square)
```

```
     # Use formula to calculate correlation
❻   numerator = n*sum_prod_x_y - sum_x*sum_y
     denominator_term1 = n*x_square_sum - squared_sum_x
     denominator_term2 = n*y_square_sum - squared_sum_y
❼   denominator = (denominator_term1*denominator_term2)**0.5
❽   correlation = numerator/denominator

     return correlation
```

The find_corr_x_y() function accepts two arguments, x and y, which are the two sets of numbers we want to calculate the correlation for. At the beginning of this function, we find the length of the lists and create a label, n, to refer to it. Next, at ❶, we have a for loop that uses the zip() function to calculate the product of the corresponding values from each list (multiplying together the first item of each list, then the second item of each list, and so on). We use the append() method to add these products to the list labeled prod.

At ❷, we calculate the sum of the products stored in prod using the sum() function. In the statements at ❸ and ❹, we calculate the sum of the numbers in x and y, respectively (once again, using the sum() function). Then, we calculate the squares of the sum of the elements in x and y, creating the labels squared_sum_x and squared_sum_y to refer to them, respectively.

In the loop starting at ❺, we calculate the square of each of the elements in x and find the sum of these squares. Then, we do the same for the elements in y. We now have all the terms we need to calculate the correlation, and we do this in the statements at ❻, ❼, and ❽. Finally, we return the correlation. Correlation is an oft-cited measure in statistical studies—in popular media and scientific articles alike. Sometimes we know ahead of time that there's a correlation, and we just want to find the strength of that correlation. We'll see an example of this in "Reading Data from a CSV File" on page 86, when we calculate the correlation between data read from a file. Other times, we might only suspect that there might be a correlation, and we must investigate the data to verify whether there actually is one (as in the following example).

High School Grades and Performance on College Admission Tests

In this section, we'll consider a fictional group of 10 students in high school and investigate whether there's a relationship between their grades in school and how they fared on their college admission tests. Table 3-2 lists the data we're going to assume for our study and base our experiments on. The "High school grades" column lists the percentile scores of the students' grades in high school, and the "College admission test scores" column shows their percentile scores on the college admission test.

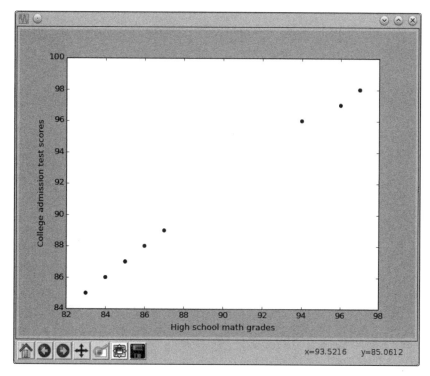

Figure 3-3: Scatter plot of high school math grades and college admission test scores

Scatter Plots

In the previous section, we saw an example of how a scatter plot can give us a first indication of the existence of any correlation between two sets of numbers. In this section, we'll see the importance of analyzing scatter plots by looking at a set of four data sets. For these data sets, conventional statistical measures all turn out to be the same, but the scatter plots of each data set reveal important differences.

First, let's go over how to create a scatter plot in Python:

```
>>> x = [1, 2, 3, 4]
>>> y = [2, 4, 6, 8]
>>> import matplotlib.pyplot as plt
❶ >>> plt.scatter(x, y)
<matplotlib.collections.PathCollection object at 0x7f351825d550>
>>> plt.show()
```

The scatter() function is used to create a scatter plot between two lists of numbers, x and y ❶. The only difference between this plot and the plots we created in Chapter 2 is that here we use the scatter() function instead of the plot() function. Once again, we have to call show() to display the plot.

To learn more about scatter plots, let's look at an important statistical study: "Graphs in Statistical Analysis" by the statistician Francis Anscombe.[1] The study considers four different data sets—referred to as *Anscombe's quartet*—with identical statistical properties: mean, variance, and correlation coefficient.

The data sets are as shown in Table 3-4 (reproduced from the original study).

Table 3-4: Anscombe's Quartet—Four Different Data Sets with Almost Identical Statistical Measures

A		B		C		D	
X1	**Y1**	**X2**	**Y2**	**X3**	**Y3**	**X4**	**Y4**
10.0	8.04	10.0	9.14	10.0	7.46	8.0	6.58
8.0	6.95	8.0	8.14	8.0	6.77	8.0	5.76
13.0	7.58	13.0	8.74	13.0	12.74	8.0	7.71
9.0	8.81	9.0	8.77	9.0	7.11	8.0	8.84
11.0	8.33	11.0	9.26	11.0	7.81	8.0	8.47
14.0	9.96	14.0	8.10	14.0	8.84	8.0	7.04
6.0	7.24	6.0	6.13	6.0	6.08	8.0	5.25
4.0	4.26	4.0	3.10	4.0	5.39	19.0	12.50
12.0	10.84	12.0	9.13	12.0	8.15	8.0	5.56
7.0	4.82	7.0	7.26	7.0	6.42	8.0	7.91
5.0	5.68	5.0	4.74	5.0	5.73	8.0	6.89

We'll refer to the pairs (X1, Y1), (X2, Y2), (X3, Y3), and (X4, Y4) as data sets A, B, C, and D, respectively. Table 3-5 presents the statistical measures of the data sets rounded off to two decimal digits.

Table 3-5: Anscombe's Quartet—Statistical Measures

	X		Y		
Data set	**Mean**	**Std. dev.**	**Mean**	**Std. dev.**	**Correlation**
A	9.00	3.32	7.50	2.03	0.82
B	9.00	3.32	7.50	2.03	0.82
C	9.00	3.32	7.50	2.03	0.82
D	9.00	3.32	7.50	2.03	0.82

The scatter plots for each data set are shown in Figure 3-4.

1. F.J. Anscombe, "Graphs in Statistical Analysis," *American Statistician* 27, no. 1 (1973): 17–21.

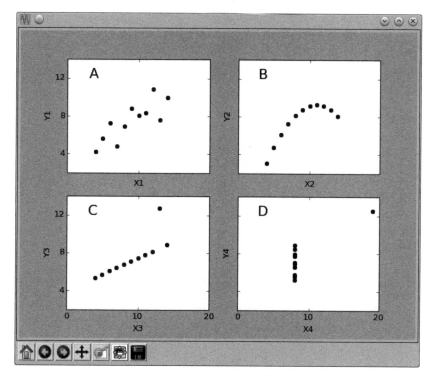

Figure 3-4: Scatter plots of Anscombe's quartet

If we look at just the traditional statistical measures (see Table 3-5)—like the mean, standard deviation, and correlation coefficient—these data sets seem nearly identical. But the scatter plots show that these data sets are actually quite different from each other. Thus, scatter plots can be an important tool and should be used alongside other statistical measures before drawing any conclusions about a data set.

Reading Data from Files

In all our programs in this chapter, the lists of numbers we used in our calculations were all explicitly written, or *hardcoded*, into the programs themselves. If you wanted to find the measures for a different data set, you'd have to enter the entire new data set in the program itself. You also know how to make programs that allow the user to enter the data as input, but with large data sets, it isn't very convenient to make the user enter long lists of numbers each time he or she uses the program.

A better alternative is to read the user data from a file. Let's see a simple example of how we can read numbers from a file and perform mathematical operations on them. First, I'll show how to read data from a simple text file with each line of the file containing a new data element. Then, I'll show you how to read from a file where the data is stored in the well-known CSV format, which will open up a lot of possibilities as there are loads of useful data sets you can download from the Internet in CSV format. (If you aren't familiar with file handling in Python, see Appendix B for a brief introduction.)

Reading Data from a Text File

Let's take a file, *mydata.txt*, with the list of donations (one per line) during period A that we considered at the beginning of this chapter:

```
100
60
70
900
100
200
500
500
503
600
1000
1200
```

The following program will read this file and print the sum of the numbers stored in the file:

```
# Find the sum of numbers stored in a file
def sum_data(filename):
    s = 0
❶    with open(filename) as f:
        for line in f:
❷            s = s + float(line)
    print('Sum of the numbers: {0}'.format(s))

if __name__ == '__main__':
    sum_data('mydata.txt')
```

The sum_data() function opens the file specified by the argument filename at ❶ and reads it line by line (f is referred to as the *file object*, and you can think of it as pointing to an opened file). At ❷, we convert each number to a floating point number using the float() function and then keep adding until we've read all the numbers. The final number, labeled s, holds the sum of the numbers, which is printed at the end of the function.

Before you run the program, you must first create a file called *mydata.txt* with the appropriate data and save it in the same directory as your program. You can create this file from IDLE itself by clicking **File ▶ New Window**, typing the numbers (one per line) in the new window, and then saving the file as *mydata.txt* in the same directory as your program. Now, if you run the program, you'll see the following output:

```
Sum of the numbers: 5733.0
```

All our programs in this chapter have assumed that the input data is available in lists. To use our earlier programs on the data from a file, we need to first create a list from that data. Once we have a list, we can use the functions we wrote earlier to calculate the corresponding statistic. The following program calculates the mean of the numbers stored in the file *mydata.txt*:

```
'''
Calculating the mean of numbers stored in a file
'''
def read_data(filename):

    numbers = []
    with open(filename) as f:
        for line in f:
❶            numbers.append(float(line))

    return numbers

def calculate_mean(numbers):
    s = sum(numbers)
    N = len(numbers)
    mean = s/N

    return mean

if __name__ == '__main__':
❷    data = read_data('mydata.txt')
    mean = calculate_mean(data)
    print('Mean: {0}'.format(mean))
```

Before we can call the calculate_mean() function, we need to read the numbers stored in the file and convert them into a list. To do this, use the read_data() function, which reads the file line by line. Instead of summing the numbers, this function converts them into floating point numbers and adds them to the list numbers ❶. The list is returned, and we refer to it by the label data ❷. We then invoke the calculate_mean() function, which returns the mean of the data. Finally, we print it.

When you run the program, you should see the following output:

```
Mean: 477.75
```

Of course, you'll see a different value for the mean if the numbers in your file are different from those in this example.

See Appendix B for hints on how you can ask the user to input the filename and then modify your program accordingly. This will allow your program's user to specify any data file.

Reading Data from a CSV File

A comma-separated value (CSV) file consists of rows and columns with the columns separated from each other by commas. You can view a CSV file using a text editor on your operating system or specialized software, such as Microsoft Excel, OpenOffice Calc, or LibreOffice Calc.

Here's a sample CSV file containing a few numbers and their squares:

```
Number,Squared
10,100
9,81
22,484
```

The first line is referred to as the *header*. In this case, it tells us that the entries in the first column of this file are numbers and those in the second column are the corresponding squares. The next three lines, or rows, contain a number and its square separated by a comma. It's possible to read the data from this file using an approach similar to what I showed for the *.txt* file. However, Python's standard library has a dedicated module (csv) for reading (and writing) CSV files, which makes things a little easier.

Save the numbers and their squares into a file, *numbers.csv*, in the same directory as your programs. The following program shows how to read this file and then create a scatter plot displaying the numbers against their squares:

```python
import csv
import matplotlib.pyplot as plt

def scatter_plot(x, y):
    plt.scatter(x, y)
    plt.xlabel('Number')
    plt.ylabel('Square')
    plt.show()

def read_csv(filename):

    numbers = []
    squared = []
    with open(filename) as f:
❶        reader = csv.reader(f)
         next(reader)
❷        for row in reader:
             numbers.append(int(row[0]))
             squared.append(int(row[1]))
         return numbers, squared

if __name__ == '__main__':
    numbers, squared = read_csv('numbers.csv')
    scatter_plot(numbers, squared)
```

The read_csv() function reads the CSV file using the reader() function defined in the csv module (which is imported at the beginning of the program). This function is called with the file object f passed to it as an argument ❶. This function then returns a *pointer* to the first line of the CSV file. We know that the first line of the file is the header, which we want to skip, so we move the pointer to the next line using the next() function. We then read every line of the file with each line referred to by the label row ❷, with row[0] referring to the first column of the data and row[1] referring to the second. For this specific file, we know that both these numbers are integers, so we use the int() function to convert these from strings to integers and to store them in two lists. The lists are then returned—one containing the numbers and the other containing the squares.

We then call the scatter_plot() function with these two lists to create the scatter plot. The find_corr_x_y() function we wrote earlier can also easily be used to find the correlation coefficient between the two sets of numbers.

Now let's try dealing with a more complex CSV file. Open *https://www .google.com/trends/correlate/* in your browser, enter any search query you wish to (for example, *summer*), and click the **Search correlations** button. You'll see that a number of results are returned under the heading "Correlated with summer," and the first result is the one with the highest correlation (the number on the immediate left of each result). Click the **Scatter plot** option above the graph to see a scatter plot with the *x*-axis labeled *summer* and the *y*-axis labeled with the top result. Ignore the exact numbers plotted on both axes as we're interested only in the correlation and the scatter plot.

A little above the scatterplot, click **Export data as CSV** and a file download will start. Save this file in the same directory as your programs.

This CSV file is slightly different from the one we saw earlier. At the beginning of the file, you'll see a number of blank lines and lines with a '#' symbol until finally you'll see the header and the data. These lines aren't useful to us—go ahead and delete them by hand using whatever software you opened the file with so that the first line of the file is the header. Also delete any blank lines at the end of the file. Now save the file. This step— where we cleaned up the file to make it easier to process with Python—is usually called *preprocessing* the data.

The header has several columns. The first contains the date of the data in each row (each row has data corresponding to the week that started on the date in this column). The second column is the search query you entered, the third column shows the search query with the *highest* correlation with your search query, and the other columns include a number of other search queries arranged in decreasing order of correlation with your entered search query. The numbers in these columns are the *z*-scores of the corresponding search queries. The *z-score* indicates the difference between the number of times a term was searched for during a specific week and the overall mean number of searches per week for that term. A positive *z*-score indicates that the number of searches was higher than the mean for that week, and a negative *z*-score indicates it was lower.

For now, let's just work with the second and the third columns. You could use the following read_csv() function to read these columns:

```
def read_csv(filename):

    with open(filename) as f:
        reader = csv.reader(f)
        next(reader)

        summer = []
        highest_correlated = []
❶       for row in reader:
            summer.append(float(row[1]))
            highest_correlated.append(float(row[2]))

    return summer, highest_correlated
```

This is pretty much like the earlier version of the read_csv function; the main change here is how we append the values to each list starting at ❶: we're now reading the second and the third members of each row, and we're storing them as floating point numbers.

The following program uses this function to calculate the correlation between the values for the search query you provided and the values for the query with the highest correlation with it. It also creates a scatter plot of these values:

```
import matplotlib.pyplot as plt
import csv

if __name__ == '__main__':
❶   summer, highest_correlated = read_csv('correlate-summer.csv')
    corr = find_corr_x_y(summer, highest_correlated)
    print('Highest correlation: {0}'.format(corr))
    scatter_plot(summer, highest_correlated)
```

Assuming that the CSV file was saved as *correlate-summer.csv*, we call the read_csv() function to read the data in the second and third columns ❶. Then, we call the find_corr_x_y() function we wrote earlier with the two lists summer and highest_correlated. It returns the correlation coefficient, which we then print. Now, we call the scatter_plot() function we wrote earlier with these two lists again. Before you can run this program, you'll need to include the definitions of the read_csv(), find_corr_x_y(), and scatter_plot() functions.

On running, you'll see that it prints the correlation coefficient and also creates a scatter plot. Both of these should be very similar to the data shown on the Google correlate website.

What You Learned

In this chapter, you learned to calculate statistical measures to describe a set of numbers and the relationships between sets of numbers. You also used graphs to aid your understanding of these measures. You learned a number of new programming tools and concepts while writing programs to calculate these measures.

Programming Challenges

Next, apply what you've learned to complete the following programming challenges.

#1: Better Correlation Coefficient–Finding Program

The find_corr_x_y() function we wrote earlier to find the correlation coefficient between two sets of numbers assumes that the two sets of numbers are the same length. Improve the function so that it first checks the length of the lists. If they're equal, only then should the function proceed with the remaining calculations; otherwise, it should print an error message that the correlation can't be found.

#2: Statistics Calculator

Implement a statistics calculator that takes a list of numbers in the file *mydata.txt* and then calculates and prints their mean, median, mode, variance, and standard deviation using the functions we wrote earlier in this chapter.

#3: Experiment with Other CSV Data

You can experiment with numerous interesting data sources freely available on the Internet. The website *http://www.quandl.com/* is one such source. For this challenge, download the following data as a CVS file from *http://www.quandl.com/WORLDBANK/USA_SP_POP_TOTL/*: the total population of the United States at the end of each year for the years 1960 to 2012. Then, calculate the mean, median, variance, and standard deviation of the *difference* in population over the years and create a graph showing these differences.

#4: Finding the Percentile

The percentile is a commonly used statistic that conveys the value below which a given percentage of observations falls. For example, if a student obtained a 95 percentile score on an exam, this means that 95 percent of the students scored less than or equal to the student's score. For another example, in the list of numbers 5, 1, 9, 3, 14, 9, and 7, the 50th percentile is 7 and the 25th percentile is 3.5, a number that is not present in the list.

There are a number of ways to find the observation corresponding to a given percentile, but here's one approach.[2]

Let's say we want to calculate the observation at percentile p:

1. In ascending order, sort the given list of numbers, which we might call data.

2. Calculate

$$i = \frac{np}{100} + 0.5,$$

where n is the number of items in data.

3. If i is an integer, data[i] is the number corresponding to percentile p.

4. If i is *not* an integer, set k equal to the integral part of i and f equal to the fractional part of i. The number (1-f)*data[k] + f*data[k+1] is the number at percentile p.

Using this approach, write a program that will take a set of numbers in a file and display the number that corresponds to a specific percentile supplied as an input to the program.

#5: Creating a Grouped Frequency Table

For this challenge, your task is to write a program that creates a grouped frequency table from a set of numbers. A grouped frequency table displays the frequency of data classified into different *classes*. For example, let's consider the scores we discussed in "Creating a Frequency Table" on page 69: 7, 8, 9, 2, 10, 9, 9, 9, 9, 4, 5, 6, 1, 5, 6, 7, 8, 6, 1, and 10. A grouped frequency table would display this data as follows:

Grade	Frequency
1–6	6
6–11	14

The table classifies the grades into two classes: 1–6 (which includes 1 but not 6) and 6–11 (which includes 6 but not 11). It displays against them the number of grades that belong to each category. Determining the number of classes and the range of numbers in each class are two key steps involved in creating this table. In this example, I've demonstrated two classes with the range of numbers in each class equally divided between the two.

---

2. See "Calculating Percentiles" by Ian Robertson (Stanford University, January 2004); *http://web.stanford.edu/class/archive/anthsci/anthsci192/anthsci192.1064/handouts/ calculating%20percentiles.pdf.*

Here's one simple approach to creating classes, which assumes the number of classes can be arbitrarily chosen:

```python
def create_classes(numbers, n):
    low = min(numbers)
    high = max(numbers)

    # Width of each class
    width = (high - low)/n
    classes = []
    a = low
    b = low + width
    classes = []
    while a < (high-width):
        classes.append((a, b))
        a = b
        b = a + width
    # The last class may be of a size that is less than width
    classes.append((a, high+1))
    return classes
```

The create_classes() function accepts two arguments: a list of numbers, numbers, and n, the number of classes to create. It'll return a list of tuples with each tuple representing a class. For example, if it's called with numbers 7, 8, 9, 2, 10, 9, 9, 9, 9, 4, 5, 6, 1, 5, 6, 7, 8, 6, 1, 10, and n = 4, it returns the following list: [(1, 3.25), (3.25, 5.5), (5.5, 7.75), (7.75, 11)]. Once you have the list, the next step is to go over each of the numbers and find out which of the returned classes it belongs to.

Your challenge is to write a program to read a list of numbers from a file and then to print the grouped frequency table, making use of the create_classes() function.

4

ALGEBRA AND SYMBOLIC MATH WITH SYMPY

The mathematical problems and solutions in our programs so far have all involved the manipulation of numbers. But there's another way math is taught, learned, and practiced, and that's in terms of symbols and the operations between them. Just think of all the *x*s and *y*s in a typical algebra problem. We refer to this type of math as *symbolic math*. I'm sure you remember those dreaded "factorize $x^3 + 3x^2 + 3x + 1$" problems in your math class. Fear no more, for in this chapter, we learn how to write programs that can solve such problems and much more. To do so, we'll use *SymPy*—a Python library that lets you write expressions containing symbols and perform operations on them. Because this is a third-party library, you'll need to install it before you can use it in your programs. The installation instructions are described in Appendix A.

Defining Symbols and Symbolic Operations

Symbols form the building blocks of symbolic math. The term *symbol* is just a general name for the *x*s, *y*s, *a*s, and *b*s you use in equations and algebraic expressions. Creating and using symbols will let us do things differently than before. Consider the following statements:

```
>>> x = 1
>>> x + x + 1
3
```

Here we create a label, x, to refer to the number 1. Then, when we write the statement x + x + 1, it's evaluated for us, and the result is 3. What if you wanted the result in terms of the symbol *x*? That is, if instead of 3, you wanted Python to tell you that the result is $2x + 1$? You couldn't just write x + x + 1 *without* the statement x = 1 because Python wouldn't know what x refers to.

SymPy lets us write programs where we can express and evaluate mathematical expressions in terms of such symbols. To use a symbol in your program, you have to create an object of the Symbol class, like this:

```
>>> from sympy import Symbol
>>> x = Symbol('x')
```

First, we import the Symbol class from the sympy library. Then, we create an object of this class passing 'x' as a parameter. Note that this 'x' is written as a string within quotes. We can now define expressions and equations in terms of this symbol. For example, here's the earlier expression:

```
>>> from sympy import Symbol
>>> x = Symbol('x')
>>> x + x + 1
2*x + 1
```

Now the result is given in terms of the symbol *x*. In the statement x = Symbol('x'), the x on the left side is the Python label. This is the same kind of label we've used before, except this time it refers to the symbol *x* instead of a number—more specifically, a Symbol object representing the symbol 'x'. This label doesn't necessarily have to match the symbol either— we could have used a label like a or var1 instead. So, it's perfectly fine to write the preceding statements as follows:

```
>>> a = Symbol('x')
>>> a + a + 1
2*x + 1
```

Using a non-matching label can be confusing, however, so I would recommend choosing a label that's the same letter as the symbol it refers to.

FINDING THE SYMBOL REPRESENTED BY
A SYMBOL OBJECT

For any Symbol object, its name attribute is a string that is the actual symbol it represents:

```
>>> x = Symbol('x')
>>> x.name
'x'
>>> a = Symbol('x')
>>> a.name
'x'
```

You can use .name on a label to retrieve the symbol that it is storing.

Just to be clear, the symbol you create has to be specified as a string. For example, you can't create the symbol *x* using x = Symbol(x)—you must define it as x = Symbol('x').

To define multiple symbols, you can either create separate Symbol objects or use the symbols() function to define them more concisely. Let's say you wanted to use three symbols—*x*, *y*, and *z*—in your program. You could define them individually, as we did earlier:

```
>>> x = Symbol('x')
>>> y = Symbol('y')
>>> z = Symbol('z')
```

But a shorter method would be to use the symbols() function to define all three at once:

```
>>> from sympy import symbols
>>> x,y,z = symbols('x,y,z')
```

First, we import the symbols() function from SymPy. Then, we call it with the three symbols we want to create, written as a string with commas separating them. After this statement is executed, x, y, and z will refer to the three symbols 'x', 'y', and 'z'.

Once you've defined symbols, you can carry out basic mathematical operations on them, using the same operators you learned in Chapter 1 (+, -, /, *, and **). For example, you might do the following:

```
>>> from sympy import Symbol
>>> x = Symbol('x')
>>> y = Symbol('y')
```

```
>>> s = x*y + x*y
>>> s
2*x*y
```

Let's see whether we can find the product of x(x + x):

```
>>> p = x*(x + x)
>>> p
2*x**2
```

SymPy will automatically make these simple addition and multiplication calculations, but if we enter a more complex expression, it will remain unchanged. Let's see what happens when we enter the expression (x + 2)*(x + 3):

```
>>> p = (x + 2)*(x + 3)
>>> p
(x + 2)*(x + 3)
```

You may have expected SymPy to multiply everything out and output x**2 + 5*x + 6. Instead, the expression was printed exactly how we entered it. SymPy automatically simplifies only the most basic of expressions and leaves it to the programmer to explicitly require simplification in cases such as the preceding one. If you want to multiply out the expression to get the expanded version, you'll have to use the expand() function, which we'll see in a moment.

Working with Expressions

Now that we know how to define our own symbolic expressions, let's learn more about using them in our programs.

Factorizing and Expanding Expressions

The factor() function decomposes an expression into its factors, and the expand() function expands an expression, expressing it as a sum of individual terms. Let's test out these functions with the basic algebraic identity $x^2 - y^2 = (x + y)(x - y)$. The left side of the identity is the expanded version, and the right side depicts the corresponding factorization. Because we have two symbols in the identity, we'll create two Symbol objects:

```
>>> from sympy import Symbol
>>> x = Symbol('x')
>>> y = Symbol('y')
```

Next, we import the `factor()` function and use it to convert the expanded version (on the left side of the identity) to the factored version (on the right side):

```
>>> from sympy import factor
>>> expr = x**2 - y**2
>>> factor(expr)
(x - y)*(x + y)
```

As expected, we get the factored version of the expression. Now let's expand the factors to get back the original expanded version:

```
>>> factors = factor(expr)
>>> expand(factors)
x**2 - y**2
```

We store the factorized expression in a new label, `factors`, and then call the `expand()` function with it. When we do this, we receive the original expression we started with. Let's try it with the more complicated identity $x^3 + 3x^2y + 3xy^2 + y^3 = (x + y)^3$:

```
>>> expr = x**3 + 3*x**2*y + 3*x*y**2 + y**3
>>> factors = factor(expr)
>>> factors
(x + y)**3

>>> expand(factors)
x**3 + 3*x**2*y + 3*x*y**2 + y**3
```

The `factor()` function is able to factorize the expression, and then the `expand()` function expands the factorized expression to return to the original expression.

If you try to factorize an expression for which there's no possible factorization, the original expression is returned by the `factor()` function. For example, see the following:

```
>>> expr = x + y + x*y
>>> factor(expr)
x*y + x + y
```

Similarly, if you pass in an expression to `expand()` that can't be expanded further, it returns the same expression.

Pretty Printing

If you want the expressions we've been working with to look a bit nicer when you print them, you can use the `pprint()` function. This function will print the expression in a way that more closely resembles how we'd normally write it on paper. For example, here's an expression:

```
>>> expr = x*x + 2*x*y + y*y
```

If we print it as we've been doing so far or use the print() function, this is how it looks:

```
>>> expr
x**2 + 2*x*y + y**2
```

Now, let's use the pprint() function to print the preceding expression:

```
>>> from sympy import pprint
>>> pprint(expr)
 2            2
x  + 2·x·y + y
```

The expression now looks much cleaner—for example, instead of having a bunch of ugly asterisks, exponents appear above the rest of the numbers.

You can also change the order of the terms when you print an expression. Consider the expression $1 + 2x + 2x^2$:

```
>>> expr = 1 + 2*x + 2*x**2
>>> pprint(expr)
   2
2·x  + 2·x + 1
```

The terms are arranged in the order of powers of x, from highest to lowest. If you want the expression in the opposite order, with the highest power of x last, you can make that happen with the init_printing() function, as follows:

```
>>> from sympy import init_printing
>>> init_printing(order='rev-lex')
>>> pprint(expr)
                2
1 + 2·x + 2·x
```

The init_printing() function is first imported and called with the keyword argument order='rev-lex'. This indicates that we want SymPy to print the expressions so that they're in *reverse lexicographical order*. In this case, the keyword argument tells Python to print the lower-power terms first.

NOTE *Although we used the init_printing() function here to set the printed order of the expressions, this function can be used in many other ways to configure how an expression is printed. For more options and to learn more about printing in SymPy, see the documentation at* http://docs.sympy.org/latest/tutorial/printing.html.

Let's apply what we've learned so far to implement a series printing program.

Printing a Series

Consider the following series:

$$x + \frac{x^2}{2} + \frac{x^3}{3} + \frac{x^4}{4} + \ldots + \frac{x^n}{n}.$$

Let's write a program that will ask a user to input a number, n, and print this series for that number. In the series, x is a symbol and n is an integer input by the program's user. The nth term in this series is given by

$$\frac{x^n}{n}.$$

We can print this series using the following program:

```
'''
Print the series:
x + x**2 + x**3 + ... + x**n
    ___  ___         ___
     2    3           n

'''

from sympy import Symbol, pprint, init_printing
def print_series(n):

    # Initialize printing system with reverse order
    init_printing(order='rev-lex')

    x = Symbol('x')
    series = x
    for i in range(2, n+1):
        series = series + (x**i)/i
    pprint(series)

if __name__ == '__main__':
    n = input('Enter the number of terms you want in the series: ')
    print_series(int(n))
```

❶ series = x
❷ for i in range(2, n+1):
❸ series = series + (x**i)/i
❹ print_series(int(n))

The print_series() function accepts an integer, n, as a parameter that is the number of terms in the series that will be printed. Note that we convert the input to an integer using the int() function when calling the function at ❹. We then call the init_printing() function to set the series to print in reverse lexicographical order.

At ❶, we create the label, series, and set its initial value as x. Then, we define a for loop that will iterate over the integers from 2 to n at ❷. Each time the loop iterates, it adds each term to series at ❸, as follows:

```
i = 2, series = x + x**2 / 2
i = 3, series = x + x**2/2 + x**3/3

--snip--
```

The value of series starts off as just plain x, but with each iteration, x**i/i gets added to the value of series until the series we want is completed. You can see SymPy addition put to good use here. Finally, the pprint() function is used to print the series.

When you run the program, it asks you to input a number and then prints the series up to that term:

```
Enter the number of terms you want in the series: 5

    x²   x³   x⁴   x⁵
x + -- + -- + -- + --
    2    3    4    5
```

Try this out with a different number of terms every time. Next, we'll see how to calculate the sum of this series for a certain value of *x*.

Substituting in Values

Let's see how we can use SymPy to plug values into an algebraic expression. This will let us calculate the value of the expression for certain values of the variables. Consider the mathematical expression $x^2 + 2xy + y^2$, which can be defined as follows:

```
>>> x = Symbol('x')
>>> y = Symbol('y')
>>> x*x + x*y + x*y + y*y
x**2 + 2*x*y + y**2
```

If you want to evaluate this expression, you can substitute numbers in for the symbols using the subs() method:

```
❶ >>> expr = x*x + x*y + x*y + y*y
  >>> res = expr.subs({x:1, y:2})
```

First, we create a new label to refer to the expression at ❶, and then we call the subs() method. The argument to the subs() method is a Python *dictionary*, which contains the two symbol labels and the numerical values we want to substitute in for each symbol. Let's check out the result:

```
>>> res
9
```

You can also express one symbol in terms of another and substitute accordingly, using the subs() method. For example, if you knew that $x = 1 - y$, here's how you could evaluate the preceding expression:

```
>>> expr.subs({x:1-y})
y**2 + 2*y*(-y + 1) + (-y + 1)**2
```

If you want the result to be simplified further—for example, if there are terms that cancel each other out, we can use SymPy's simplify() function, as follows:

```
❶ >>> expr_subs = expr.subs({x:1-y})
  >>> from sympy import simplify
❷ >>> simplify(expr_subs)
  1
```

At ❶, we create a new label, expr_subs, to refer to the result of substituting $x = 1 - y$ in the expression. We then import the simplify() function from SymPy and call it at ❷. The result turns out to be 1 because the other terms of the expression cancel each other.

Although there was a simplified version of the expression in the preceding example, you had to ask SymPy to simplify it using the simplify() function. Once again, this is because SymPy won't do any simplification without being asked to.

The simplify() function can also simplify complicated expressions, such as those including logarithms and trigonometric functions, but we won't get into that here.

Calculating the Value of a Series

Let's revisit the series-printing program. In addition to printing the series, we want our program to be able to find the value of the series for a particular value of x. That is, our program will now take two inputs from the user—the number of terms in the series and the value of x for which the value of the series will be calculated. Then, the program will output both the series and the sum. The following program extends the series printing program to include these enhancements:

```
'''
Print the series:
x + x**2 + x**3 + ... + x**n
    ___   ___         ___
     2     3           n
'''

from sympy import Symbol, pprint, init_printing
def print_series(n, x_value):

    # Initialize printing system with reverse order
    init_printing(order='rev-lex')

    x = Symbol('x')
    series = x
    for i in range(2, n+1):
        series = series + (x**i)/i

    pprint(series)

    # Evaluate the series at x_value
❶    series_value = series.subs({x:x_value})
    print('Value of the series at {0}: {1}'.format(x_value, series_value))


if __name__ == '__main__':
    n = input('Enter the number of terms you want in the series: ')
❷    x_value = input('Enter the value of x at which you want to evaluate the series: ')

    print_series(int(n), float(x_value))
```

The print_series() function now takes an additional argument, x_value, which is the value of x for which the series should be evaluated. At ❶, we use the subs() method to perform the evaluation and the label series_value to refer to the result. In the next line, we display the result.

The additional input statement at ❷ asks the user to enter the value of x using the label x_value to refer to it. Before we call the print_series() function, we convert this value into its floating point equivalent using the float() function.

If you execute the program now, it will ask you for the two inputs and print out the series and the series value:

```
Enter the number of terms you want in the series: 5
Enter the value of x at which you want to evaluate the series: 1.2

    x²   x³   x⁴   x⁵
x + -- + -- + -- + --
    2    3    4    5
Value of the series at 1.2: 3.51206400000000
```

In this sample run, we ask for five terms in the series, with x set to 1.2, and the program prints and evaluates the series.

Converting Strings to Mathematical Expressions

So far, we've been writing out individual expressions each time we want to do something with them. However, what if you wanted to write a more general program that could manipulate any expression provided by the user? For that, we need a way to convert a user's input, which is a string, into something we can perform mathematical operations on. SymPy's sympify() function helps us do exactly that. The function is so called because it converts the string into a SymPy object that makes it possible to apply SymPy's functions to the input. Let's see an example:

```
❶ >>> from sympy import sympify
   >>> expr = input('Enter a mathematical expression: ')
   Enter a mathematical expression: x**2 + 3*x + x**3 + 2*x
❷ >>> expr = sympify(expr)
```

We first import the sympify() function at ❶. We then use the input() function to ask for a mathematical expression as input, using the label expr to refer to it. Next, we call the sympify() function with expr as its argument at ❷ and use the same label to refer to the converted expression.

You can perform various operations on this expression. For example, let's try multiplying the expression by 2:

```
>>> 2*expr
2*x**3 + 2*x**2 + 10*x
```

What happens when the user supplies an invalid expression? Let's see:

```
>>> expr = input('Enter a mathematical expression: ')
Enter a mathematical expression: x**2 + 3*x + x**3 + 2x
>>> expr = sympify(expr)
Traceback (most recent call last):
  File "<pyshell#146>", line 1, in <module>
    expr = sympify(expr)
  File "/usr/lib/python3.3/site-packages/sympy/core/sympify.py", line 180, in sympify
    raise SympifyError('could not parse %r' % a)
sympy.core.sympify.SympifyError: SympifyError: "could not parse 'x**2 + 3*x + x**3 + 2x'"
```

The last line tells us that `sympify()` isn't able to convert the supplied input expression. Because this user didn't add an operator between 2 and x, SymPy doesn't understand what it means. Your program should expect such invalid input and print an error message if it comes up. Let's see how we can do that by catching the `SympifyError` exception:

```
>>> from sympy import sympify
>>> from sympy.core.sympify import SympifyError
>>> expr = input('Enter a mathematical expression: ')
Enter a mathematical expression: x**2 + 3*x + x**3 + 2x
>>> try:
    expr = sympify(expr)
except SympifyError:
    print('Invalid input')

Invalid input
```

The two changes in the preceding program are that we import the `SympifyError` exception class from the `sympy.core.sympify` module and call the `sympify()` function in a try...except block. Now if there's a `SympifyError` exception, an error message is printed.

Expression Multiplier

Let's apply the `sympify()` function to write a program that calculates the product of two expressions:

```
'''
Product of two expressions
'''

from sympy import expand, sympify
from sympy.core.sympify import SympifyError

def product(expr1, expr2):
    prod = expand(expr1*expr2)
    print(prod)

if __name__=='__main__':
❶    expr1 = input('Enter the first expression: ')
❷    expr2 = input('Enter the second expression: ')

    try:
        expr1 = sympify(expr1)
        expr2 = sympify(expr2)
    except SympifyError:
        print('Invalid input')
    else:
❸        product(expr1, expr2)
```

At ❶ and ❷, we ask the user to enter the two expressions. Then, we convert them into a form understood by SymPy using the `sympify()` function in a try...except block. If the conversion succeeds (indicated by the else block), we call the product() function at ❸. In this function, we calculate the product of the two expressions and print it. Note how we use the expand() function to print the product so that all its terms are expressed as a sum of its constituent terms.

Here's a sample execution of the program:

```
Enter the first expression: x**2 + x*2 + x
Enter the second expression: x**3 + x*3 + x
x**5 + 3*x**4 + 4*x**3 + 12*x**2
```

The last line displays the product of the two expressions. The input can also have more than one symbol in any of the expressions:

```
Enter the first expression: x*y+x
Enter the second expression: x*x+y
x**3*y + x**3 + x*y**2 + x*y
```

Solving Equations

SymPy's solve() function can be used to find solutions to equations. When you input an expression with a symbol representing a variable, such as x, solve() calculates the value of that symbol. This function always makes its calculation by assuming the expression you enter is equal to zero—that is, it prints the value that, when substituted for the symbol, makes the entire expression equal zero. Let's start with the simple equation $x - 5 = 7$. If we want to use solve() to find the value of x, we first have to make one side of the equation equal zero ($x - 5 - 7 = 0$). Then, we're ready to use solve(), as follows:

```
>>> from sympy import Symbol, solve
>>> x = Symbol('x')
>>> expr = x - 5 - 7
>>> solve(expr)
[12]
```

When we use solve(), it calculates the value of 'x' as 12 because that's the value that makes the expression ($x - 5 - 7$) equal to zero.

Note that the result 12 is returned in a list. An equation can have multiple solutions—for example, a quadratic equation has two solutions. In that case, the list will have all the solutions as its members. You can also ask the solve() function to return the result so that each member is dictionary instead. Each dictionary is composed of the symbol (variable name) and its value (the solution). This is especially useful when solving simultaneous equations where we have more than one variable to solve for because when the solution is returned as a dictionary, we know which solution corresponds to which variable.

Solving Quadratic Equations

In Chapter 1, we found the roots of the quadratic equation $ax^2 + bx + c = 0$ by writing the formulas for the two roots and then substituting the values of the constants a, b, and c. Now, we'll learn how we can use SymPy's solve() function to find the roots without needing to write out the formulas. Let's see an example:

```
❶ >>> from sympy import solve
   >>> x = Symbol('x')
❷ >>> expr = x**2 + 5*x + 4
❸ >>> solve(expr, dict=True)
❹ [{x: -4}, {x: -1}]
```

The solve() function is first imported at ❶. We then define a symbol, x, and an expression corresponding to the quadratic equation, x**2 + 5*x + 4, at ❷. Then, we call the solve() function with the preceding expression at ❸. The second argument to the solve() function (dict=True) specifies that we want the result to be returned as a list of Python dictionaries.

Each solution in the returned list is a dictionary using the symbol as a key matched with its corresponding value. If the solution is empty, an empty list will be returned. The roots of the preceding equation are –4 and –1, as you can see at ❹.

We found out in the first chapter that the roots of the equation

$$x^2 + x + 1 = 0$$

are complex numbers. Let's attempt to find those using solve():

```
>>> x=Symbol('x')
>>> expr = x**2 + x + 1
>>> solve(expr, dict=True)
[{x: -1/2 - sqrt(3)*I/2}, {x: -1/2 + sqrt(3)*I/2}]
```

Both the roots are imaginary, as expected with the imaginary component indicated by the I symbol.

Solving for One Variable in Terms of Others

In addition to finding the roots of equations, we can take advantage of symbolic math to use the solve() function to express one variable in an equation in terms of the others. Let's take a look at finding the roots for the generic quadratic equation $ax^2 + bx + c = 0$. To do so, we'll define x and three additional symbols—a, b, and c, which correspond to the three constants:

```
>>> x = Symbol('x')
>>> a = Symbol('a')
>>> b = Symbol('b')
>>> c = Symbol('c')
```

Next, we write the expression corresponding to the equation and use the solve() function on it:

```
>>> expr = a*x*x + b*x + c
>>> solve(expr, x, dict=True)
[{x: (-b + sqrt(-4*a*c + b**2))/(2*a)}, {x: -(b + sqrt(-4*a*c + b**2))/(2*a)}]
```

Here, we have to include an additional argument, x, to the solve() function. Because there's more than one symbol in the equation, we need to tell solve() which symbol it should solve for, which is what we indicate by passing in x as the second argument. As we'd expect, solve() prints the quadratic formula: the generic formula for finding the value(s) of x in a polynomial expression.

To be clear, when we use solve() on an equation with more than one symbol, we specify the symbol to solve for as the second argument (and now the third argument specifies how we want the results to be returned).

Next, let's consider an example from physics. According to one of the equations of motion, the distance traveled by a body moving with a constant acceleration a, with an initial velocity u, in time t, is given by

$$s = ut + \frac{1}{2}at^2.$$

Given u and a, however, if you wanted to find the time required to travel a given distance, s, you'd have to first express t in terms of the other variables. Here's how you could do that using SymPy's solve() function:

```
>>> from sympy import Symbol, solve, pprint
>>> s = Symbol('s')
>>> u = Symbol('u')
>>> t = Symbol('t')
>>> a = Symbol('a')
>>> expr = u*t + (1/2)*a*t*t - s
>>> t_expr = solve(expr,t, dict=True)
>>> pprint(t_expr)
```

The result looks like this:

$$\left[\left\{t: \frac{-u + \sqrt{2.0 \cdot a \cdot s + u^2}}{a}\right\}, \left\{t: \frac{-\left(u + \sqrt{2.0 \cdot a \cdot s + u^2}\right)}{a}\right\}\right]$$

Now that we have the expression for t (referred to by the label t_expr), we can use the subs() method to replace the values of s, u, and a to find the two possible values of t.

Solving a System of Linear Equations

Consider the following two equations:

$$2x + 3y = 6$$

$$3x + 2y = 12$$

Say we want to find the pair of values (x, y) that satisfies both the equations. We can use the solve() function to find the solution for a system of equations like this one.

First, we define the two symbols and create the two equations:

```
>>> x = Symbol('x')
>>> y = Symbol('y')
>>> expr1 = 2*x + 3*y - 6
>>> expr2 = 3*x + 2*y - 12
```

The two equations are defined by the expressions expr1 and expr2, respectively. Note how we've rearranged the expressions so they both equal zero (we moved the right side of the given equations to the left side). To find the solution, we call the solve() function with the two expressions forming a tuple:

```
>>> solve((expr1, expr2), dict=True)
[{y: -6/5, x: 24/5}]
```

As I mentioned earlier, getting the solution back as a dictionary is useful here. We can see that the value of x is 24/5 and the value of y is −6/5. Let's verify whether the solution we got really satisfies the equations. To do so, we'll first create a label, soln, to refer to the solution we got and then use the subs() method to substitute the corresponding values of x and y in the two expressions:

```
>>> soln = solve((expr1, expr2), dict=True)
>>> soln = soln[0]
>>> expr1.subs({x:soln[x], y:soln[y]})
0
>>> expr2.subs({x:soln[x], y:soln[y]})
0
```

The result of substituting the values of x and y corresponding to the solution in the two expressions is zero.

Plotting Using SymPy

In Chapter 2, we learned to make graphs where we explicitly specified the numbers we wanted to plot. For example, to plot the graph of the gravitational force against the distance between two bodies, you had to calculate the gravitational force for each distance value and supply the lists of distances and forces to matplotlib. With SymPy, on the other hand, you can just tell SymPy the equation of the line you want to plot, and the graph will be created for you. Let's plot a line whose equation is given by $y = 2x + 3$:

```
>>> from sympy.plotting import plot
>>> from sympy import Symbol
>>> x = Symbol('x')
>>> plot(2*x+3)
```

All we had to do was import plot and Symbol from sympy.plotting, create a symbol, x, and call the plot() function with the expression 2*x+3. SymPy takes care of everything else and plots the graph of the function, as shown in Figure 4-1.

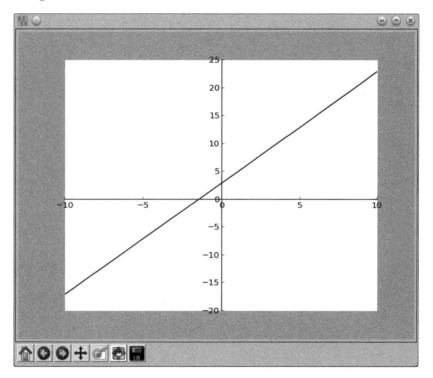

Figure 4-1: Plot of the line y = 2x + 3

The graph shows that a default range of *x* values was automatically chosen: −10 to 10. You may notice that the graph window looks very similar to those you saw in Chapters 2 and 3. That's because SymPy uses matplotlib behind the scenes to draw the graphs. Also note that we didn't have to call the show() function to show the graphs because this is done automatically by SymPy.

Now, let's say that you wanted to limit the values of 'x' in the preceding graph to lie in the range –5 to 5 (instead of –10 to 10). You'd do that as follows:

```
>>> plot((2*x + 3), (x, -5, 5))
```

Here, a tuple consisting of the symbol, the lower bound, and the upper bound of the range—(x, -5, 5)—is specified as the second argument to the plot() function. Now, the graph displays only the values of *y* corresponding to the values of *x* between –5 and 5 (see Figure 4-2).

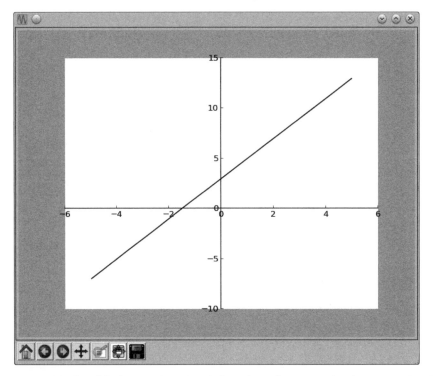

Figure 4-2: Plot of the line y = 2x + 3 with the values of x restricted to the range –5 to 5

You can use other keyword arguments in the plot() function, such as title to enter a title or xlabel and ylabel to label the *x*-axis and the *y*-axis, respectively. The following plot() function specifies the preceding three keyword arguments (see the corresponding graph in Figure 4-3):

```
>>> plot(2*x + 3, (x, -5, 5), title='A Line', xlabel='x', ylabel='2x+3')
```

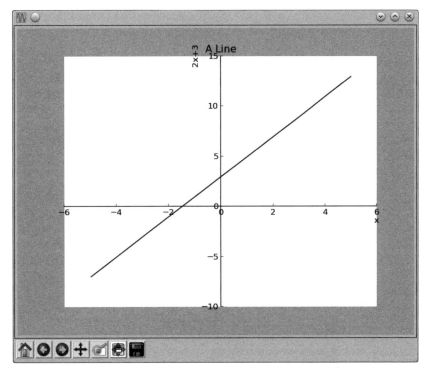

Figure 4-3: Plot of the line y = 2x + 3 with the range of x and other attributes specified

The plot shown in Figure 4-3 now has a title and labels on the *x*-axis and the *y*-axis. You can specify a number of other keyword arguments to the plot() function to customize the behavior of the function as well as the graph itself. The show keyword argument allows us to specify whether we want the graph to be displayed. Passing show=False will cause the graph to not be displayed when you call the plot() function:

```
>>> p = plot(2*x + 3, (x, -5, 5), title='A Line', xlabel='x', ylabel='2x+3', show=False)
```

You will see that no graph is shown. The label p refers to the plot that is created, so you can now call p.show() to display the graph. You can also save the graph as an image file using the save() method, as follows:

```
>>> p.save('line.png')
```

This will save the plot to a file *line.png* in the current directory.

Plotting Expressions Input by the User

The expression that you pass to the plot() function must be expressed in terms of *x* only. For example, earlier we plotted $y = 2x + 3$, which we entered to the plot function as simply $2x + 3$. If the expression were not originally in this form, we'd have to rewrite it. Of course, we could do this manually, outside the program. But what if you want to write a program that allows its users to graph any expression? If the user enters an expression in the form of $2x + 3y - 6$, say, we have to first convert it. The solve() function will help us here. Let's see an example:

```
>>> expr = input('Enter an expression: ')
Enter an expression: 2*x + 3*y - 6
❶ >>> expr = sympify(expr)
❷ >>> y = Symbol('y')
>>> solve(expr, y)
❸ [-2*x/3 + 2]
```

At ❶, we use the sympify() function to convert the input expression to a SymPy object. At ❷, we create a Symbol object to represent 'y' so that we can tell SymPy which variable we want to solve the equation for. Then we solve the expression to find y in terms of x by specifying y as the second argument to the solve() function. At ❸, this returns the equation in terms of x, which is what we need for plotting.

Notice that this final expression is stored in a list, so before we can use it, we'll have to extract it from the list:

```
>>> solutions = solve(expr, 'y')
❹ >>> expr_y = solutions[0]
>>> expr_y
-2*x/3 + 2
```

We create a label, solutions, to refer to the result returned by the solve() function, which is a list with only one item. Then, we extract that item at ❹. Now, we can call the plot() function to graph the expression. The next listing shows a full graph-drawing program:

```
'''
Plot the graph of an input expression
'''

from sympy import Symbol, sympify, solve
from sympy.plotting import plot

def plot_expression(expr):

    y = Symbol('y')
    solutions = solve(expr, y)
    expr_y = solutions[0]
    plot(expr_y)

if __name__=='__main__':

    expr = input('Enter your expression in terms of x and y: ')
```

```
try:
    expr = sympify(expr)
except SympifyError:
    print('Invalid input')
else:
    plot_expression(expr)
```

Note that the preceding program includes a try...except block to check for invalid input, as we've done with sympify() earlier. When you run the program, it asks you to input an expression, and it will create the corresponding graph.

Plotting Multiple Functions

You can enter multiple expressions when calling the SymPy plot function to plot more than one expression on the same graph. For example, the following code plots two lines at once (see Figure 4-4):

```
>>> from sympy.plotting import plot
>>> from sympy import Symbol
>>> x = Symbol('x')
>>> plot(2*x+3, 3*x+1)
```

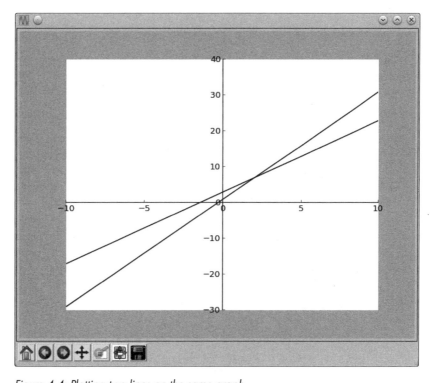

Figure 4-4: Plotting two lines on the same graph

This example brings out another difference between plotting in matplotlib and in SymPy. Here, using SymPy, both lines are the same color, whereas matplotlib would have automatically made the lines different colors. To set different colors for each line with SymPy, we'll need to perform some extra steps, as shown in the following code, which also adds a legend to the graph:

```
>>> from sympy.plotting import plot
>>> from sympy import Symbol
>>> x = Symbol('x')
❶ >>> p = plot(2*x+3, 3*x+1, legend=True, show=False)
❷ >>> p[0].line_color = 'b'
❸ >>> p[1].line_color = 'r'
>>> p.show()
```

At ❶, we call the plot() function with the equations for the two lines but pass two additional keyword arguments—legend and show. By setting the legend argument to True, we add a legend to the graph, as we saw in Chapter 2. Note, however, that the text that appears in the legend will match the expressions you plotted—you can't specify any other text. We also set show=False because we want to set the color of the lines before we draw the graph. The statement at ❷, p[0], refers to the first line, $2x + 3$, and we set its attribute line_color to 'b', meaning that we want this line to be blue. Similarly, we set the color of the second plot to red using the string 'r' ❸. Finally, we call the show() to display the graph (see Figure 4-5).

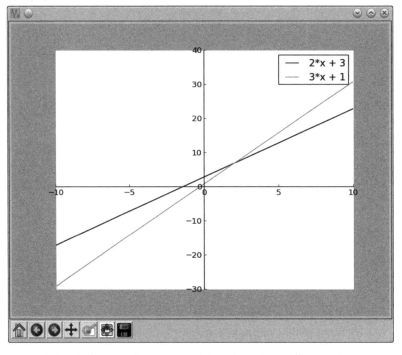

Figure 4-5: Plot of the two lines with each line drawn in a different color

In addition to red and blue, you can plot the lines in green, cyan, magenta, yellow, black, and white (using the first letter of the color in each case).

What You Learned

In this chapter, you learned the basics of symbolic math using SymPy. You learned about declaring symbols, constructing expressions using symbols and mathematical operators, solving equations, and plotting graphs. You will be learning more features of SymPy in later chapters.

Programming Challenges

Here are a few programming challenges that should help you further apply what you've learned. You can find sample solutions at *http://www.nostarch .com/doingmathwithpython/*.

#1: Factor Finder

You learned about the factor() function, which prints the factors of an expression. Now that you know how your program can handle expressions input by a user, write a program that will ask the user to input an expression, calculate its factors, and print them. Your program should be able to handle invalid input by making use of exception handling.

#2: Graphical Equation Solver

Earlier, you learned how to write a program that prompts the user to input an expression such as $3x + 2y - 6$ and create the corresponding graph. Write a program that asks the user for two expressions and then graphs them both, as follows:

```
>>> expr1 = input('Enter your first expression in terms of x and y: ')
>>> expr2 = input('Enter your second expression in terms of x and y: ')
```

Now, expr1 and expr2 will store the two expressions input by the user. You should convert both of these into SymPy objects using the sympify() step in a try...except block.

All you need to do from here is plot these two expressions instead of one.

Once you've completed this, enhance your program to print the solution—the pair of x and y values that satisfies both equations. This will also be the spot where the two lines on the graph intersect. (Hint: Refer to how we used the solve() function earlier to find the solution of a system of two linear equations.)

#3: Summing a Series

We saw how to find the sum of a series in "Printing a Series" on page 99. There, we manually added the terms of the series by looping over all the terms. Here's a snippet from that program:

```
for i in range(2, n+1):
    series = series + (x**i)/i
```

SymPy's summation() function can be directly used to find such summations. The following example prints the sum of the first five terms of the series we considered earlier:

```
>>> from sympy import Symbol, summation, pprint
>>> x = Symbol('x')
>>> n = Symbol('n')
❶ >>> s = summation(x**n/n, (n, 1, 5))
>>> pprint(s)
 5    4    3    2
x    x    x    x
-- + -- + -- + -- + x
5    4    3    2
```

We call the summation() function at ❶, with the first argument being the nth term of the series and the second argument being a tuple that states the range of n. We want the sum of the first five terms here, so the second argument is (n, 1, 5).

Once you have the sum, you can use the subs() method to substitute a value for x to find the numerical value of the sum:

```
>>> s.subs({x:1.2})
3.51206400000000
```

Your challenge is to write a program that's capable of finding the sum of an arbitrary series when you supply the nth term of the series and the number of terms in it. Here's an example of how the program would work:

```
Enter the nth term: a+(n-1)*d
Enter the number of terms: 3
3·a + 3·d
```

In this example, the nth term supplied is that of an *arithmetic progression*. Starting with a and d as the *common difference*, the number of terms up to which the sum is to be calculated is 3. The sum turns out to be 3a + 3d, which agrees with the known formula for the same.

#4: Solving Single-Variable Inequalities

You've seen how to solve an equation using SymPy's solve() function. But SymPy is also capable of solving single-variable inequalities, such as $x + 5 > 3$ and $\sin x - 0.6 > 0$. That is, SymPy can solve relations besides equality, like >, <, and so on. For this challenge, create a function, isolve(), that will take any inequality, solve it, and then return the solution.

First, let's learn about the SymPy functions that will help you implement this. The inequality-solving functions are available as three separate functions for polynomial, rational, and all other inequalities. We'll need to pick the right function to solve various inequalities, or we'll get an error.

A *polynomial* is an algebraic expression consisting of a variable and coefficients and involving only the operations of addition, subtraction, and multiplication and only positive powers of the variable. An example of a polynomial inequality is $x^2 + 4 < 0$.

To solve a polynomial inequality, use the solve_poly_inequality() function:

```
>>> from sympy import Poly, Symbol, solve_poly_inequality
>>> x = Symbol('x')
❶ >>> ineq_obj = -x**2 + 4 < 0
❷ >>> lhs = ineq_obj.lhs
❸ >>> p = Poly(lhs, x)
❹ >>> rel = ineq_obj.rel_op
>>> solve_poly_inequality(p, rel)
[(-oo, -2), (2, oo)]
```

First, create the expression representing an inequality, $-x^2 + 4 < 0$, at ❶ and refer to this expression with the label ineq_obj. Then, extract the left side of the inequality—that is, the algebraic expression $-x^2 + 4$—using the lhs attribute at ❷. Next, create a Poly object at ❸ to represent the polynomial we extracted at ❷. The second argument passed when creating the object is the symbol object that represents the variable, x. At ❹, extract the relational operator from the inequality object using the rel attribute. Finally, call the solve_poly_inequality() function with the polynomial object, p, and rel as the two arguments. The program returns the solution as a list of tuples, with each tuple representing a solution for the inequality as the lower limit and the upper limit of the range of numbers. For this inequality, the solution is all numbers less than −2 and all numbers greater than 2.

A *rational expression* is an algebraic expression in which the numerator and denominator are both polynomials. Here's an example of a rational inequality:

$$\frac{x-1}{x+2} > 0$$

For rational inequalities, use the solve_rational_inequalities() function:

```
>>> from sympy import Symbol, Poly, solve_rational_inequalities
>>> x = Symbol('x')
❶ >>> ineq_obj = ((x-1)/(x+2)) > 0
>>> lhs = ineq_obj.lhs
❷ >>> numer, denom = lhs.as_numer_denom()
>>> p1 = Poly(numer)
>>> p2 = Poly(denom)
>>> rel = ineq_obj.rel_op
❸ >>> solve_rational_inequalities([[((p1, p2), rel)]])
(-oo, -2) U (1, oo)
```

Create an inequality object representing our example rational inequality at ❶ and then extract the rational expression using the lhs attribute. Separate out the numerator and the denominator into the labels numer and denom using the as_numer_denom() method at ❷, which returns a tuple with the numerator and denominator as the two members. Then, create two polynomial objects, p1 and p2, representing the numerator and denominator, respectively. Retrieve the relational operator and call the solve_rational_inequalities() function, passing it the two polynomial objects—p1 and p2—and the relational operator.

The program returns the solution (-oo, -2) U (1, oo), where U denotes that the solution is a *union* of the two *sets* of solutions consisting of all numbers less than –2 and all numbers greater than 1. (We'll learn about sets in Chapter 5.)

Finally, $\sin x - 0.6 > 0$ is an example of an inequality that belongs to neither the polynomial nor rational expression categories. If you have such an inequality to solve, use the solve_univariate_inequality() function:

```
>>> from sympy import Symbol, solve, solve_univariate_inequality, sin
>>> x = Symbol('x')
>>> ineq_obj = sin(x) - 0.6 > 0
>>> solve_univariate_inequality(ineq_obj, x, relational=False)
(0.643501108793284, 2.49809154479651)
```

Create an inequality object representing the inequality sin(x) - 0.6 > 0 and then call the solve_univariate_inequality() function with the first two arguments as the inequality object, ineq_obj, and the symbol object, x. The keyword argument relational=False specifies to the function that we want the solution to be returned as a *set*. The solution for this inequality turns out to be all numbers lying between the first and second members of the tuple the program returns.

Hints: Handy Functions

Now remember—your challenge is (1) to create a function, isolve(), that will take any inequality and (2) to choose one of the appropriate functions discussed in this section to solve it and return the solution. The following hints may be useful to implement this function.

The is_polynomial() method can be used to check whether an expression is a polynomial or not:

```
>>> x = Symbol('x')
>>> expr = x**2 - 4
>>> expr.is_polynomial()
True
>>> expr = 2*sin(x) + 3
>>> expr.is_polynomial()
False
```

The is_rational_function() can be used to check whether an expression is a rational expression:

```
>>> expr = (2+x)/(3+x)
>>> expr.is_rational_function()
True
>>> expr = 2+x
>>> expr.is_rational_function()
True
>>> expr = 2+sin(x)
>>> expr.is_rational_function()
False
```

The sympify() function can convert an inequality expressed as a string to an inequality object:

```
>>> from sympy import sympify
>>> sympify('x+3>0')
x + 3 > 0
```

When you run your program, it should ask the user to input an inequality expression and print back the solution.

5

PLAYING WITH
SETS AND PROBABILITY

In this chapter, we'll start by learning how we can make our programs understand and manipulate sets of numbers. We'll then see how sets can help us understand basic concepts in probability. Finally, we'll learn about generating random numbers to simulate random events. Let's get started!

What's a Set?

A *set* is a collection of distinct objects, often called *elements* or *members*. Two characteristics of a set make it different from just any collection of objects. A set is "well defined," meaning the question "Is a particular object in this collection?" always has a clear yes or no answer, usually based on a rule or some given criteria. The second characteristic is that no two members of a set are the same. A set can contain anything—numbers, people, things, words, and so on.

Let's walk through some basic properties of sets as we learn how to work with sets in Python using SymPy.

Set Construction

In mathematical notation, you represent a set by writing the set members enclosed in curly brackets. For example, {2, 4, 6} represents a set with 2, 4, and 6 as its members. To create a set in Python, we can use the FiniteSet class from the sympy package, as follows:

```
>>> from sympy import FiniteSet
>>> s = FiniteSet(2, 4, 6)
>>> s
{2, 4, 6}
```

Here, we first import the FiniteSet class from SymPy and then create an object of this class by passing in the set members as arguments. We assign the label s to the set we just created.

We can store different types of numbers—including integers, floating point numbers, and fractions—in the same set:

```
>>> from sympy import FiniteSet
>>> from fractions import Fraction
>>> s = FiniteSet(1, 1.5, Fraction(1, 5))
>>> s
{1/5, 1, 1.5}
```

The *cardinality* of a set is the number of members in the set, which you can find by using the len() function:

```
>>> s = FiniteSet(1, 1.5, 3)
>>> len(s)
3
```

Checking Whether a Number Is in a Set

To check whether a number is a member of an existing set, use the in operator. This operator asks Python, "Is this number in this set?" It returns True if the number belongs to the set and False if it doesn't. If, for example, we wanted to check whether 4 was in the previous set, we'd do the following:

```
>>> 4 in s
False
```

Because 4 is not present in the set, the operator returns False.

Creating an Empty Set

If you want to make an *empty set,* which is a set that doesn't have any elements or members, create a FiniteSet object without passing any arguments. The result is an EmptySet object:

```
>>> s = FiniteSet()
>>> s
EmptySet()
```

Creating Sets from Lists or Tuples

You can also create a set by passing in a list or a tuple of set members as an argument to FiniteSet:

```
>>> members = [1, 2, 3]
>>> s = FiniteSet(*members)
>>> s
{1, 2, 3}
```

Here, instead of passing in the set members directly to FiniteSet, we first stored them in a list, which we called members. Then, we passed the list to FiniteSet using this special Python syntax, which basically translates to creating a FiniteSet object that passes the list members as separate arguments and not as a list. That is, this approach to creating a FiniteSet object is equivalent to FiniteSet(1, 2, 3). We will make use of this syntax when the set members are computed at runtime.

Set Repetition and Order

Sets in Python (like mathematical sets) ignore any repeats of a member, and they don't keep track of the order of set members. For example, if you create a set from a list that has multiple instances of a number, the number is added to the set only once, and the other instances are discarded:

```
>>> from sympy import FiniteSet
>>> members = [1, 2, 3, 2]
>>> FiniteSet(*members)
{1, 2, 3}
```

Here, even though we passed in a list that had two instances of the number 2, the number 2 appears only once in the set created from that list.

In Python lists and tuples, each element is stored in a particular order, but the same is not always true for sets. For example, we can print out each member of a set by iterating through it as follows:

```
>>> from sympy import FiniteSet
>>> s = FiniteSet(1, 2, 3)
>>> for member in s:
        print(member)
```

```
2
1
3
```

When you run this code, the elements could be printed in any possible order. This is because of how sets are stored by Python—it keeps track of what members are in the set, but it doesn't keep track of any particular order for those members.

Let's see another example. Two sets are *equal* when they have the same elements. In Python, you can use the equality operator, ==, to check whether two sets are equal:

```
>>> from sympy import FiniteSet
>>> s = FiniteSet(3, 4, 5)
>>> t = FiniteSet(5, 4, 3)
>>> s == t
True
```

Although the members of these two sets appear in different orders, the sets are still equal.

Subsets, Supersets, and Power Sets

A set, s, is a *subset* of another set, *t*, if all the members of *s* are also members of *t*. For example, the set {1} is a subset of the set {1, 2}. You can check whether a set is a subset of another set using the is_subset() method:

```
>>> s = FiniteSet(1)
>>> t = FiniteSet(1,2)
>>> s.is_subset(t)
True
>>> t.is_subset(s)
False
```

Note that an empty set is a subset of every set. Also, any set is a subset of itself, as you can see in the following:

```
>>> s.is_subset(s)
True
>>> t.is_subset(t)
True
```

Similarly, a set, *t*, is said to be a *superset* of another set, *s*, if *t* contains all of the members contained in *s*. You can check whether one set is a superset of another using the is_superset() method:

```
>>> s.is_superset(t)
False
>>> t.is_superset(s)
True
```

The *power set* of a set, *s*, is the set of all possible subsets of *s*. Any set, *s*, has precisely $2^{|s|}$ subsets, where |s| is the cardinality of the set. For example, the set {1, 2, 3} has a cardinality of 3, so it has 2^3 or 8 subsets: {} (the empty set), {1}, {2}, {3}, {1, 2}, {2, 3}, {1, 3}, and {1, 2, 3}.

The set of all these subsets form the power set, and we can find the power set using the powerset() method:

```
>>> s = FiniteSet(1, 2, 3)
>>> ps = s.powerset()
>>> ps
{{1}, {1, 2}, {1, 3}, {1, 2, 3}, {2}, {2, 3}, {3}, EmptySet()}
```

As the power set is a set itself, you can find its cardinality using the len() function:

```
>>> len(ps)
8
```

The cardinality of the power set is $2^{|s|}$, which is $2^3 = 8$.

Based on our definition of a subset, any two sets with the exact same members would be subsets as well as supersets of each other. By contrast, a set, *s*, is a *proper subset* of *t* only if all the members of *s* are also in *t* and *t* has at least one member that is not in *s*. So if *s* = {1, 2, 3}, it's only a proper subset of *t* if *t* contains 1, 2, and 3 plus at least one more member. This would also mean that *t* is a *proper superset* of *s*. You can use the is_proper_subset() method and the is_proper_superset() method to check for these relationships:

```
>>> from sympy import FiniteSet
>>> s = FiniteSet(1, 2, 3)
>>> t = FiniteSet(1, 2, 3)
>>> s.is_proper_subset(t)
False
>>> t.is_proper_superset(s)
False
```

Now, if we re-create the set t to include another member, s will be considered a proper subset of t and t a proper superset of s:

```
>>> t = FiniteSet(1, 2, 3, 4)
>>> s.is_proper_subset(t)
True
>>> t.is_proper_superset(s)
True
```

Set Operations

Set operations such as union, intersection, and the Cartesian product allow you to combine sets in certain methodical ways. These set operations are extremely useful in real-world problem-solving situations when we have to consider multiple sets together. Later in this chapter, we'll see how to use these operations to apply a formula to multiple sets of data and calculate the probabilities of random events.

Union and Intersection

The *union* of two sets is a set that contains all of the *distinct* members of the two sets. In set theory, we use the symbol $\cup$ to refer to the union operation. For example, {1, 2} $\cup$ {2, 3} will result in a new set, {1, 2, 3}. In SymPy, the union of these two sets can be created using the union() method:

```
>>> from sympy import FiniteSet
>>> s = FiniteSet(1, 2, 3)
>>> t = FiniteSet(2, 4, 6)
>>> s.union(t)
{1, 2, 3, 4, 6}
```

We find the union of s and t by applying the union method to s and passing in t as an argument. The result is a third set with all the distinct members of the two sets. In other words, each member of this third set is a member of one or both of the first two sets.

The *intersection* of two sets creates a new set from the elements common to both sets. For example, the intersection of the sets {1, 2} and {2, 3} will result in a new set with the only common element, {2}. Mathematically, this operation is written as {1, 2} ∩ {2, 3}.

In SymPy, use the intersect() method to find the intersection:

```
>>> s = FiniteSet(1, 2)
>>> t = FiniteSet(2, 3)
>>> s.intersect(t)
{2}
```

Whereas the union operation finds members that are in one set *or* another, the intersection operation finds elements that are present in both. Both of these operations can also be applied to more than two sets. For example, here's how you'd find the union of three sets:

```
>>> from sympy import FiniteSet
>>> s = FiniteSet(1, 2, 3)
>>> t = FiniteSet(2, 4, 6)
>>> u = FiniteSet(3, 5, 7)
>>> s.union(t).union(u)
{1, 2, 3, 4, 5, 6, 7}
```

Similarly, here's how you'd find the intersection of three sets:

```
>>> s.intersect(t).intersect(u)
EmptySet()
```

The intersection of the sets s, t, and u turns out to be an empty set because there are no elements that all three sets share.

Cartesian Product

The *Cartesian product* of two sets creates a set that consists of all possible pairs made by taking an element from each set. For example, the Cartesian product of the sets {1, 2} and {3, 4} is {(1, 3), (1, 4), (2, 3), (2, 4)}. In SymPy, you can find the Cartesian product of two sets by simply using the multiplication operator:

```
>>> from sympy import FiniteSet
>>> s = FiniteSet(1, 2)
>>> t = FiniteSet(3, 4)
>>> p = s*t
>>> p
{1, 2} x {3, 4}
```

This takes the Cartesian product of the sets s and t and stores it as p. To actually see each pair in that Cartesian product, we can iterate through and print them out as follows:

```
>>> for elem in p:
        print(elem)
(1, 3)
(1, 4)
(2, 3)
(2, 4)
```

Each element of the product is a tuple consisting of a member from the first set and a member from the second set.

The cardinality of the Cartesian product is the product of the cardinality of the individual sets. We can demonstrate this in Python:

```
>>> len(p) ==  len(s)*len(t)
True
```

If we apply the exponential operator (**) to a set, we get the Cartesian product of that set times itself the specified number of times.

```
>>> from sympy import FiniteSet
>>> s = FiniteSet(1, 2)
>>> p = s**3
>>> p
{1, 2} x {1, 2} x {1, 2}
```

Here, for example, we raised the set s to the power of 3. Because we're taking the Cartesian product of three sets, this gives us a set of all possible triplets that contain a member of each set:

```
>>> for elem in p:
        print(elem)
(1, 1, 1)
(1, 1, 2)
(1, 2, 1)
(1, 2, 2)
(2, 1, 1)
(2, 1, 2)
(2, 2, 1)
(2, 2, 2)
```

Finding the Cartesian product of sets is useful for finding all possible combinations of the set members, which we'll explore next.

Applying a Formula to Multiple Sets of Variables

Consider a simple pendulum of length L. The *time period*, T, of this pendulum—that is, the amount of time it takes for the pendulum to complete one full swing—is given by the formula

$$T = 2\pi \sqrt{\frac{L}{g}}.$$

Here, π is the mathematical constant, *pi*, and g is the local gravitational acceleration, which is around 9.8 m/s^2 on Earth. Because π and g are constants, the length, L, is the only variable on the right side of the equation that doesn't have a constant value.

If you wanted to see how the time period of a simple pendulum varies with its length, you'd assume different values for the length and measure the corresponding time period at each of these values using the formula. A typical high school experiment is to compare the time period you get using the preceding formula, which is the theoretical result, to the one you measure in the laboratory, which is the experimental result. For example, let's choose five different values: 15, 18, 21, 22.5, and 25 (all expressed in centimeters). With Python, we can write a quick program that'll speed through the calculations for the theoretical results:

```
  from sympy import FiniteSet, pi
❶ def time_period(length):
      g = 9.8
      T = 2*pi*(length/g)**0.5
      return T

  if __name__ == '__main__':
❷     L = FiniteSet(15, 18, 21, 22.5, 25)
      for l in L:
❸         t = time_period(l/100)
          print('Length: {0} cm Time Period: {1:.3f} s'. format(float(l), float(t)))
```

We first define the function time_period at ❶. This function applies the formula shown earlier to a given length, which is passed in as length. Then, our program defines a set of lengths at ❷ and applies the time_period function to each value at ❸. Notice that when we pass in the length values to time_period, we divide them by 100. This operation converts the lengths from centimeters to meters so that they match the unit of gravitational acceleration, which is expressed in units of meters/second^2. Finally, we print the calculated time period. When you run the program, you'll see the following output:

```
Length: 15.0 cm Time Period: 0.777 s
Length: 18.0 cm Time Period: 0.852 s
Length: 21.0 cm Time Period: 0.920 s
Length: 22.5 cm Time Period: 0.952 s
Length: 25.0 cm Time Period: 1.004 s
```

Different Gravity, Different Results

Now, imagine we conducted this experiment in three different places—my current location, Brisbane, Australia; the North Pole; and the equator. The force of gravity varies slightly depending on the latitude of your location: it's a bit lower (approximately 9.78 m/s^2) at the equator and higher (9.83 m/s^2) at the North Pole. This means we can treat the force of gravity as a variable in our formula, rather than a constant, and calculate results for three different values of gravitational acceleration: {9.8, 9.78, 9.83}.

If we want to calculate the period of a pendulum for each of our five lengths at each of these three locations, a systematic way to work out all of these combinations of the values is to take the Cartesian product, as shown in the following program:

```
from sympy import FiniteSet, pi

def time_period(length, g):

    T = 2*pi*(length/g)**0.5
    return T

if __name__ == '__main__':

    L = FiniteSet(15, 18, 21, 22.5, 25)
    g_values = FiniteSet(9.8, 9.78, 9.83)
❶  print('{0:^15}{1:^15}{2:^15}'.format('Length(cm)', 'Gravity(m/s^2)', 'Time  Period(s)'))
❷  for elem in L*g_values:
❸      l = elem[0]
❹      g = elem[1]
        t = time_period(l/100, g)

❺      print('{0:^15}{1:^15}{2:^15.3f}'.format(float(l), float(g), float(t)))
```

At ❷, we take the Cartesian product of our two sets of variables, L and g_values, and we iterate through each resulting combination of values to calculate the time period. Each combination is represented as a tuple, and for each tuple, we extract the first value, the length, at ❸ and the second value, the gravity, at ❹. Then, just as before, we call the time_period() function with these two labels as parameters, and we print the values of length (l), gravity (g), and the corresponding time period (T).

The output is presented in a table to make it easy to follow. The table is formatted by the print statements at ❶ and ❺. The format string {0:^15}{1:^15}{2:^15.3f} creates three fields, each 15 spaces wide, and the ^ symbol centers each entry in each field. In the last field of the print statement at ❺, '.3f' limits the number of digits after the decimal point to three.

When you run the program, you'll see the following output:

Length(cm)	Gravity(m/s^2)	Time Period(s)
15.0	9.78	0.778
15.0	9.8	0.777
15.0	9.83	0.776

18.0	9.78	0.852
18.0	9.8	0.852
18.0	9.83	0.850
21.0	9.78	0.921
21.0	9.8	0.920
21.0	9.83	0.918
22.5	9.78	0.953
22.5	9.8	0.952
22.5	9.83	0.951
25.0	9.78	1.005
25.0	9.8	1.004
25.0	9.83	1.002

This experiment presents a simple scenario where you need all possible combinations of the elements of multiple sets (or a group of numbers). In this type of situation, the Cartesian product is exactly what you need.

Probability

Sets allow us to reason about the basic concepts of probability. We'll begin with a few definitions:

Experiment   The *experiment* is simply the test we want to perform. We perform the test because we're interested in the probability of each possible outcome. Rolling a die, flipping a coin, and pulling a card from a deck of cards are all examples of experiments. A single run of an experiment is referred to as a *trial*.

Sample space   All possible outcomes of an experiment form a set known as the *sample space*, which we'll usually call *S* in our formulas. For example, when a six-sided die is rolled once, the sample space is {1, 2, 3, 4, 5, 6}.

Event   An *event* is a set of outcomes that we want to calculate the probability of and that form a *subset* of the sample space. For example, we might want to know the probability of a particular outcome, like rolling a 3, or the probability of a set of multiple outcomes, such as rolling an even number (either 2, 4, or 6). We'll use the letter *E* in our formulas to stand for an event.

If there's a *uniform distribution*—that is, if each outcome in the sample space is equally likely to occur—then the probability of an event, $P(E)$, occurring is calculated using the following formula (I'll talk about nonuniform distributions a bit later in this chapter):

$$P(E) = \frac{n(E)}{n(S)}.$$

Here, $n(E)$ and $n(S)$ are the cardinality of the sets E, the event, and S, the sample space, respectively. The value of $P(E)$ ranges from 0 to 1, with higher values indicating a higher chance of the event happening.

We can apply this formula to a die roll to calculate the probability of a particular roll—say, 3:

$$S = \{1, 2, 3, 4, 5, 6\}$$

$$E = \{3\}$$

$$n(S) = 6$$

$$n(E) = 1$$

$$P(E) = \frac{1}{6}$$

This confirms what was obvious all along: the probability of a particular die roll is 1/6. You could easily do this calculation in your head, but we can use this formula to write the following function in Python that calculates the probability of any event, event, in any sample space, space:

```
def probability(space, event):
    return len(event)/len(space)
```

In this function, the two arguments space and event—the sample space and event—need not be sets created using FiniteSet. They can also be lists or, for that matter, any other Python object that supports the len() function.

Using this function, let's write a program to find the probability of a prime number appearing when a 20-sided die is rolled:

```
  def probability(space, event):
      return len(event)/len(space)

❶ def check_prime(number):
      if number != 1:
          for factor in range(2, number):
              if number % factor == 0:
                  return False
      else:
          return False
      return True

  if __name__ == '__main__':
❷     space = FiniteSet(*range(1, 21))
      primes = []
      for num in s:
❸         if check_prime(num):
              primes.append(num)
❹     event= FiniteSet(*primes)
      p = probability(space, event)

      print('Sample space: {0}'.format(space))
      print('Event: {0}'.format(event))
      print('Probability of rolling a prime: {0:.5f}'.format(p))
```

We first create a set representing the sample space, space, using the range() function at ❷. To create the event set, we need to find the prime numbers from the sample space, so we define a function, check_prime(), at ❶. This function takes an integer and checks to see whether it's divisible (with no remainder) by any number between 2 and itself. If so, it returns False. Because a prime number is only divisible by 1 and itself, this function returns True if an integer is prime and False otherwise.

We call this function for each of the numbers in the sample space at ❸ and add the prime numbers to a list, primes. Then, we create our event set, event, from this list at ❹. Finally, we call the probability() function we created earlier. We get the following output when we run the program:

```
Sample space: {1, 2, 3, ..., 18, 19, 20}
Event: {2, 3, 5, 7, 11, 13, 17, 19}
Probability of rolling a prime: 0.40000
```

Here, $n(E) = 8$ and $n(S) = 20$, so the probability, P, is 0.4.

In our 20-sided die program, we really didn't need to create the sets; instead, we could have called the probability() function with the sample space and events as lists:

```
if __name__ == '__main__':
    space = range(1, 21)
    primes = []
    for num in space:
        if check_prime(num):
            primes.append(num)
    p = probability(space, primes)
```

The probability() function works equally well in this case.

Probability of Event A or Event B

Let's say we're interested in two possible events, and we want to find the probability of *either* one of them happening. For example, going back to a simple die roll, let's consider the following two events:

A = The number is a prime number.

B = The number is odd.

As it was earlier, the sample space, S, is {1, 2, 3, 4, 5, 6}. Event A can be represented as the subset {2, 3, 5}, the set of prime numbers in the sample space, and event B can be represented as {1, 3, 5}, the odd numbers in the sample space. To calculate the probability of either set of outcomes, we can find the probability of the *union* of the two sets. In our notation, we could say:

$$E = \{2, 3, 5\} \cup \{1, 3, 5\} = \{1, 2, 3, 5\}$$

$$P(E) = \frac{n(E)}{n(S)} = \frac{4}{6} = \frac{2}{3}$$

Now let's perform this calculation in Python:

```
>>> from sympy import FiniteSet
>>> s = FiniteSet(1, 2, 3, 4, 5, 6)
>>> a = FiniteSet(2, 3, 5)
>>> b = FiniteSet(1, 3, 5)
❶ >>> e = a.union(b)
>>> len(e)/len(s)
0.6666666666666666
```

We first create a set, s, representing the sample space, followed by the two sets a and b. Then, at ❶, we use the union() method to find the event set, e. Finally, we calculate the probability of the union of the two sets using the earlier formula.

Probability of Event A and Event B

Say you have two events in mind and you want to calculate the chances of *both* of them happening—for example, the chances that a die roll is both prime and odd. To determine this, you calculate the probability of the intersection of the two event sets:

$$E = A \cap B = \{2, 3, 5\} \cap \{1, 3, 5\} = \{3, 5\}$$

We can calculate the probability of both A and B happening by using the intersect() method, which is similar to what we did in the previous case:

```
>>> from sympy import FiniteSet
>>> s = FiniteSet(1, 2, 3, 4, 5, 6)
>>> a = FiniteSet(2, 3, 5)
>>> b = FiniteSet(1, 3, 5)
>>> e = a.intersect(b)
>>> len(e)/len(s)
0.3333333333333333
```

Generating Random Numbers

Probability concepts allow us to reason about and calculate the chance of an event happening. To actually simulate such events—like a simple dice game—using computer programs, we need a way to generate random numbers.

Simulating a Die Roll

To simulate a six-sided die roll, we need a way to generate a random integer between 1 and 6. The random module in Python's standard library provides us with various functions to generate random numbers. Two functions that we'll use in this chapter are the randint() function, which generates a random integer in a given range, and the random() function, which generates a floating point number between 0 and 1. Let's see a quick example of how the randint() function works:

```
>>> import random
>>> random.randint(1, 6)
4
```

The randint() function takes two integers as arguments and returns a random integer that falls between these two numbers (both inclusive). In this example, we passed in the range (1, 6), and it returned the number 4, but if we call it again, we'll very likely get a different number:

```
>>> random.randint(1, 6)
6
```

Calling the randint() function allows us to simulate the roll of our virtual die. Every time we call this function, we're going to get a number between 1 and 6, just as we would if we were rolling a six-sided die. Note that randint() expects you to supply the lower number first, so randint(6, 1) isn't valid.

Can You Roll That Score?

Our next program will simulate a simple die-rolling game, where we keep rolling the six-sided die until we've rolled a total of 20:

```
'''
Roll a die until the total score is 20
'''

import matplotlib.pyplot as plt
import random

target_score = 20

def roll():
    return random.randint(1, 6)

if __name__ == '__main__':
    score = 0
    num_rolls = 0
 ❶  while score < target_score:
        die_roll = roll()
        num_rolls += 1
        print('Rolled: {0}'.format(die_roll))
        score += die_roll

    print('Score of {0} reached in {1} rolls'.format(score, num_rolls))
```

First, we define the same roll() function we created earlier. Then, we use a while loop at ❶ to call this function, keep track of the number of rolls, print the current roll, and add up the total score. The loop repeats until the score reaches 20, and then the program prints the total score and number of rolls.

Here's a sample run:

```
Rolled: 6
Rolled: 2
Rolled: 5
Rolled: 1
Rolled: 3
Rolled: 4
Score of 21 reached in 6 rolls
```

If you run the program several times, you'll notice that the number of rolls it takes to reach 20 varies.

Is the Target Score Possible?

Our next program is similar, but it'll tell us whether a certain target score is reachable within a maximum number of rolls:

```
from sympy import FiniteSet
import random

def find_prob(target_score, max_rolls):

    die_sides = FiniteSet(1, 2, 3, 4, 5, 6)
    # Sample space
❶   s = die_sides**max_rolls
    # Find the event set
    if max_rolls > 1:
        success_rolls = []
❷       for elem in s:
            if sum(elem) >= target_score:
                success_rolls.append(elem)
    else:
        if target_score > 6:
❸           success_rolls = []
        else:
            success_rolls = []
            for roll in die_sides:
❹               if roll >= target_score:
                    success_rolls.append(roll)
❺   e = FiniteSet(*success_rolls)
    # Calculate the probability of reaching target score
    return len(e)/len(s)

if __name__ == '__main__':

    target_score = int(input('Enter the target score: '))
    max_rolls    = int(input('Enter the maximum number of rolls allowed: '))
```

```
p = find_prob(target_score, max_rolls)
print('Probability: {0:.5f}'.format(p))
```

When you run this program, it asks for the target score and the maximum number of allowed rolls as input, and then it prints out the probability of achieving that.

Here are two sample executions:

```
Enter the target score: 25
Enter the maximum number of rolls allowed: 4
Probability:  0.00000

Enter the target score: 25
Enter the maximum number of rolls allowed: 5
Probability:  0.03241
```

Let's understand the workings of the `find_prob()` function, which performs the probability calculation. The sample space here is the Cartesian product, $die\_sides^{max\_rolls}$ ❶, where `die_sides` is the set $\{1, 2, 3, 4, 5, 6\}$, representing the numbers on a six-sided die, and `max_rolls` is the maximum number of die rolls allowed.

The event set is all the sets in the sample space that help us reach this target score. There are two cases here: when the number of turns left is greater than 1 and when we're in the last turn. For the first case, at ❷, we iterate over each of the tuples in the Cartesian product and add the ones that add up to or exceed `target_score` in the `success_rolls` list. The second case is special: our sample space is just the set $\{1, 2, 3, 4, 5, 6\}$, and we have only one throw of the die left. If the value of the target score is greater than 6, it isn't possible to achieve it, and we set `success_rolls` to be an empty list at ❸. If however, the `target_score` is less than or equal to 6, we iterate over each possible roll and add the ones that are greater than or equal to the value of `target_score` at ❹.

At ❺, we calculate the event set, e, from the `success_rolls` list that we constructed earlier and then return the probability of reaching the target score.

Nonuniform Random Numbers

Our discussions of probability have so far assumed that each of the outcomes in the sample space is equally likely. The `random.randint()` function, for example, returns an integer in the specified range assuming that each integer is *equally likely*. We refer to such probability as *uniform probability* and to random numbers generated by the `randint()` function as *uniform random numbers*. Let's say, however, that we want to simulate a biased coin toss—a loaded coin for which heads is twice as likely to occur as tails. We'd then need a way to generate *nonuniform* random numbers.

Before we write the program to do so, we'll review the idea behind it.

Consider a number line with a length of 1 and with two equally divided intervals, as shown in Figure 5-1.

Figure 5-1: A number line with a length of 1 divided into two equal intervals corresponding to the probability of heads or tails on a coin toss

We'll refer to this line as the *probability number line*, with each division representing an equally possible outcome—for example, heads or tails upon a fair coin toss. Now, in Figure 5-2, consider a different version of this number line.

Figure 5-2: A number line with a length of 1 divided into two unequal intervals corresponding to the probability of heads or tails on a biased coin toss

Here, the division corresponding to heads is 2/3 of the total length and the division corresponding to tails is 1/3. This represents the situation of a coin that's likely to turn up heads in 2/3 of tosses and tails only in 1/3 of tosses. The following Python function will simulate such a coin toss, considering this unequal probability of heads or tails appearing:

```
import random

def toss():
    # 0 -> Heads, 1-> Tails
❶   if random.random() < 2/3:
        return 0
    else:
        return 1
```

We assume that the function returns 0 to indicate heads and 1 to indicate tails, and then we generate a random number between 0 and 1 at ❶ using the `random.random()` function. If the generated number is less than 2/3—the probability of flipping heads with our biased coin—the program returns 0; otherwise it returns 1 (tails).

We'll now see how we can extrapolate the preceding function to simulate a nonuniform event with multiple possible outcomes. Let's consider a fictional ATM that dispenses a \$5, \$10, \$20, or \$50 bill when its button is pressed. The different denominations have varying probabilities of being dispensed, as shown in Figure 5-3.

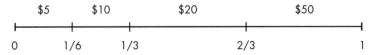

Figure 5-3: A number line with a length of 1 divided into four intervals of different lengths corresponding to the probability of dispensing bills of different denominations

Here, the probability of a $5 bill or $10 bill being dispensed is 1/6, and the probability of a $20 bill or $50 bill being dispensed is 1/3.

We create a list to store the rolling sum of the probabilities, and then we generate a random number between 0 and 1. We start from the left end of the list that stores the sum and return the first index of this list for which the corresponding sum is lesser than or equal to the random number generated. The get_index() function implements this idea:

```
'''
Simulate a fictional ATM that dispenses dollar bills
of various denominations with varying probability
'''

import random

def get_index(probability):
    c_probability = 0
❶  sum_probability = []
    for p in probability:
        c_probability += p
        sum_probability.append(c_probability)
❷  r = random.random()
    for index, sp in enumerate(sum_probability):
❸      if r <= sp:
            return index
❹  return len(probability)-1

def dispense():

    dollar_bills = [5, 10, 20, 50]
    probability = [1/6, 1/6, 1/3, 2/3]
    bill_index = get_index(probability)
    return dollar_bills[bill_index]
```

We call the get_index() function with a list containing the probability that the event in the corresponding position is expected to occur. We then, at ❶, construct the list sum_probability, where the ith element is the sum of the first i elements in the list probability. That is, the first element in sum_probability is equal to the first element in probability, the second element is equal to the sum of the first two elements in probability, and so on. At ❷, a random number between 0 and 1 is generated using the label r. Next, at ❸, we traverse through sum_probability and return the index of the first element that exceeds r.

The last line of the function, at ❹, takes care of a special case best illustrated through an example. Consider a list of three events with percentages of occurrence each expressed as 0.33. In this case, the list sum_probability would look like [0.33, 0.66, 0.99]. Now, consider that the random number generated, r, is 0.99314. For this value of r, we want the last element in the list of events to be chosen. You may argue that this isn't exactly right because the last event has a higher than 33 percent chance of being selected. As per the condition at ❸, there's no element in sum_probability that's greater than r; hence, the function wouldn't return any index at all. The statement at ❹ takes care of this and returns the last index.

If you call the dispense() function to simulate a large number of dollar bills disbursed by the ATM, you'll see that the ratio of the number of times each bill appears closely obeys the probability specified. We'll find this technique useful when creating *fractals* in the next chapter.

What You Learned

In this chapter, you started by learning how to represent a set in Python. Then, we discussed the various set concepts and you learned about the union, the intersection, and the Cartesian product of sets. You applied some of the set concepts to explore the basics of probability and, finally, learned how to simulate uniform and nonuniform random events in your programs.

Programming Challenges

Next, you have a few programming challenges to solve that'll give you the opportunity to apply what you've learned in this chapter.

#1: Using Venn Diagrams to Visualize Relationships Between Sets

A *Venn diagram* is an easy way to see the relationship between sets graphically. It tells us how many elements are common between the two sets, how many elements are only in one set, and how many elements are in neither set. Consider a set, A, that represents the set of positive odd numbers less than 20—that is, $A = \{1, 3, 5, 7, 9, 11, 13, 15, 17, 19\}$—and consider another set, B, that represents the set of prime numbers less than 20—that is, $B = \{2, 3, 5, 7, 11, 13, 17, 19\}$. We can draw Venn diagrams with Python using the matplotlib_venn package (see Appendix A for installation instructions for this package). Once you've installed it, you can draw the Venn diagram as follows:

```
'''
Draw a Venn diagram for two sets
'''

from matplotlib_venn import venn2
import matplotlib.pyplot as plt
from sympy import FiniteSet

def draw_venn(sets):

    venn2(subsets=sets)
    plt.show()
```

```
if __name__ == '__main__':

    s1 = FiniteSet(1, 3, 5, 7, 9, 11, 13, 15, 17, 19)
    s2 = FiniteSet(2, 3, 5, 7, 11, 13, 17, 19)

    draw_venn([s1, s2])
```

Once we import all the required modules and functions (the venn2() function, matplotlib.pyplot, and the FiniteSet class), all we have to do is create the two sets and then call the venn2() function, using the subsets keyword argument to specify the sets as a tuple.

Figure 5-4 shows the Venn diagram created by the preceding program. The sets *A* and *B* share seven common elements, so 7 is written in the common area. Each of the sets also has unique elements, so the number of unique elements—3 and 1, respectively—is written in the individual areas. The labels below the two sets are shown as *A* and *B*. You can specify your own labels using the set_labels keyword argument:

```
>>> venn2(subsets=(a,b), set_labels=('S', 'T'))
```

This would change the set labels to S and T.

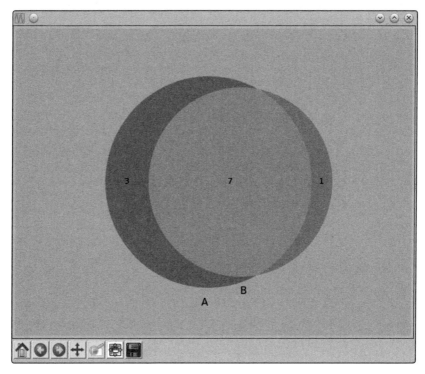

Figure 5-4: Venn diagram showing the relationship between two sets, A and B

For your challenge, imagine you've created an online questionnaire asking your classmates the following question: *Do you play football, another sport, or no sports?* Once you have the results, create a CSV file, *sports.csv*, as follows:

```
StudentID,Football,Others
1,1,0
2,1,1
3,0,1
--snip--
```

Create 20 such rows for the 20 students in your class. The first column is the student ID (the survey isn't anonymous), the second column has a 1 if the student has marked "football" as the sport they love to play, and the third column has a 1 if the student plays any other sport or none at all. Write a program to create a Venn diagram to depict the summarized results of the survey, as shown in Figure 5-5.

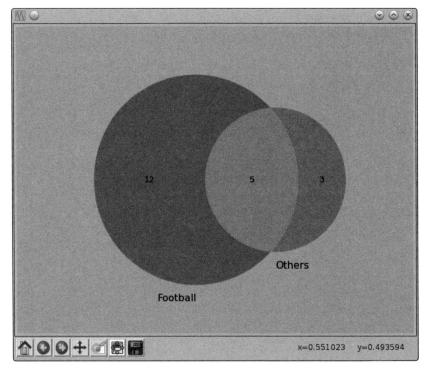

Figure 5-5: A Venn diagram showing the number of students who love to play football and the number who love to play other sports

Depending on the data in the *sports.csv* file you created, the numbers in each set will vary. The following function reads a CSV file and returns two lists corresponding to the IDs of those students who play football and other sports:

```
def read_csv(filename):
    football = []
    others = []
    with open(filename) as f:
        reader = csv.reader(f)
        next(reader)
        for row in reader:
            if row[1] == '1':
                football.append(row[0])
            if row[2] == '1':
                others.append(row[0])

    return football, others
```

#2: Law of Large Numbers

We've referred to a die roll and coin toss as two examples of random events that we can simulate using random numbers. We've used the term *event* to refer to a certain number showing up on a die roll or to heads or tails showing up on a coin toss, with each event having an associated probability value. In probability, a *random variable*—usually denoted by *X*—describes an event. For example, *X* = 1 describes the event of 1 appearing upon a die roll, and *P*(*X* = 1) describes the associated probability. There are two kinds of random variables: (1) *discrete* random variables, which take only integral values and are the only kind of random variables we see in this chapter, and (2) *continuous* random variables, which—as the name suggests—can take any real value.

The *expectation, E*, of a discrete random variable is the equivalent of the average or mean that we learned about in Chapter 3. The expectation can be calculated as follows:

$$E = x_1 P(x_1) + x_2 P(x_2) + x_3 P(x_3) + \ldots + x_n P(x_n)$$

Thus, for a six-sided die, the *expected value* of a die roll can be calculated like this:

```
>>> e = 1*(1/6) + 2*(1/6) + 3*(1/6) + 4*(1/6) + 5*(1/6) + 6*(1/6)
>>> e
3.5
```

According to the *law of large numbers*, the average value of results over multiple trials approaches the expected value as the number of trials increases. Your challenge in this task is to verify this law when rolling a six-sided die for the following number of trials: 100, 1000, 10000, 100000, and 500000. Here's an expected sample run of your complete program:

```
Expected value: 3.5
Trials: 100 Trial average 3.39
Trials: 1000 Trial average 3.576
Trials: 10000 Trial average 3.5054
Trials: 100000 Trial average 3.50201
Trials: 500000 Trial average 3.495568
```

#3: How Many Tosses Before You Run Out of Money?

Let's consider a simple game played with a fair coin toss. A player wins $1 for heads and loses $1.50 for tails. The game is over when the player's balance reaches $0. Given a certain starting amount specified by the user as input, your challenge is to write a program that simulates this game. Assume there's an unlimited cash reserve with the computer—your opponent here. Here's a possible game play session:

```
Enter your starting amount: 10
Tails! Current amount: 8.5
Tails! Current amount: 7.0
Tails! Current amount: 5.5
Tails! Current amount: 4.0
Tails! Current amount: 2.5
Heads! Current amount: 3.5
Tails! Current amount: 2.0
Tails! Current amount: 0.5
Tails! Current amount: -1.0
Game over :( Current amount: -1.0. Coin tosses: 9
```

#4: Shuffling a Deck of Cards

Consider a standard deck of 52 playing cards. Your challenge here is to write a program to simulate the shuffling of this deck. To keep the implementation simple, I suggest you use the integers 1, 2, 3, . . . , 52 to represent the deck. Every time you run the program, it should output a shuffled deck—in this case, a shuffled list of integers.

Here's a possible output of your program:

```
[3, 9, 21, 50, 32, 4, 20, 52, 7, 13, 41, 25, 49, 36, 23, 45, 1, 22, 40, 19, 2,
35, 28, 30, 39, 44, 29, 38, 48, 16, 15, 18, 46, 31, 14, 33, 10, 6, 24, 5, 43,
47, 11, 34, 37, 27, 8, 17, 51, 12, 42, 26]
```

The `random` module in Python's standard library has a function, `shuffle()`, for this exact operation:

```
>>> import random
>>> x = [1, 2, 3, 4]
❶ >>> random.shuffle(x)
>>> x
[4, 2, 1, 3]
```

Create a list, x, consisting of the numbers [1, 2, 3, 4]. Then, call the `shuffle()` function ❶, passing this list as an argument. You'll see that the numbers in x have been shuffled. Note that the list is shuffled "in place." That is, the original order is lost.

But what if you wanted to use this program in a card game? There, it's not enough to simply output the shuffled list of integers. You'll also need a way to map back the integers to the specific suit and rank of each card. One way you might do this is to create a Python class to represent a single card:

```
class Card:
    def __init__(self, suit, rank):
        self.suit = suit
        self.rank = rank
```

To represent the ace of clubs, create a card object, `card1 = Card('clubs', 'ace')`. Then, do the same for all the other cards. Next, create a list consisting of each of the card objects and shuffle this list. The result will be a shuffled deck of cards where you also know the suit and rank of each card. Output of the program should look something like this:

```
10 of spades
6 of clubs
jack of spades
9 of spades
```

#5: Estimating the Area of a Circle

Consider a dartboard with a circle of radius r inscribed in a square with side $2r$. Now let's say you start throwing a large number of darts at it. Some of these will hit the board within the circle—let's say, N—and others outside it—let's say, M. If we consider the fraction of darts that land inside the circle,

$$f = \frac{N}{N + M},$$

then the value of $f \times A$, where A is the area of the square, would roughly be equal to the area of the circle (see Figure 5-6). The darts are represented by the small circular dots in the figure. We shall refer to the value of $f \times A$ as the estimated area. The actual area is, of course, πr^2.

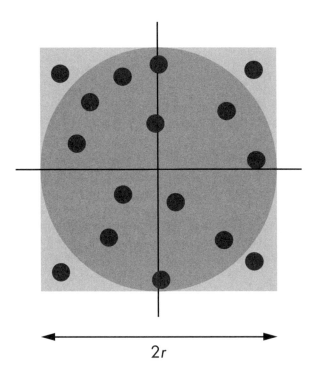

Figure 5-6: A circle of radius r inscribed in a square board with
side 2r. The dots represent darts randomly thrown at the board.

As part of this challenge, write a program that will find the estimated
area of a circle, given any radius, using this approach. The program should
print the estimated area of the circle for three different values of the number of darts: 10^3, 10^5, and 10^6. That's a lot of darts! You'll see that increasing the number of darts brings the estimated area close to the actual area.
Here's a sample output of the completed solution:

```
Radius: 2
Area: 12.566370614359172, Estimated (1000 darts): 12.576
Area: 12.566370614359172, Estimated (100000 darts): 12.58176
Area: 12.566370614359172, Estimated (1000000 darts): 12.560128
```

The dart throw can be simulated by a call to the random.uniform(a, b)
function, which will return a random number between *a* and *b*. In this case,
use the values *a* = 0, *b* = 2*r* (the side of the square).

Estimating the Value of Pi

Consider Figure 5-6 once again. The area of the square is $4r^2$, and the area of the inscribed circle is πr^2. If we divide the area of the circle by the area of the square, we get $\pi/4$. The fraction f that we calculated earlier,

$$f = \frac{N}{N + M},$$

is thus an approximation of $\pi/4$, which in turn means that the value of

$$4\frac{N}{N + M}$$

should be close to the value of π. Your next challenge is to write a program that will estimate the value of π assuming any value for the radius. As you increase the number of darts, the estimated value of π should get close to the known value of the constant.

6

DRAWING GEOMETRIC SHAPES
AND FRACTALS

In this chapter, we'll start by learning about patches in matplotlib that allow us to draw geometric shapes, such as circles, triangles, and polygons. We'll then learn about matplotlib's animation support and write a program to animate a projectile's trajectory. In the final section, we'll learn how to draw *fractals*—complex geometric shapes created by the repeated applications of simple geometric transformations. Let's get started!

Drawing Geometric Shapes with Matplotlib's Patches

In matplotlib, *patches* allow us to draw geometric shapes, each of which we refer to as a *patch*. You can specify, for example, a circle's radius and center in order to add the corresponding circle to your plot. This is quite different from how we've used matplotlib so far, which has been to supply the *x*- and *y*-coordinates of the points to plot. Before we can write a program to make use of the patches feature, however, we'll need to understand a little bit more about how a matplotlib plot is created. Consider the following program, which plots the points (1, 1), (2, 2), and (3, 3) using matplotlib:

```
>>> import matplotlib.pyplot as plt
>>> x = [1, 2, 3]
>>> y = [1, 2, 3]
>>> plt.plot(x, y)
[<matplotlib.lines.Line2D object at 0x7fe822d67a20>]
>>> plt.show()
```

This program creates a matplotlib window that shows a line passing through the given points. Under the hood, when the plt.plot() function is called, a Figure object is created, within which the axes are created, and finally the data is plotted within the axes (see Figure 6-1).[1]

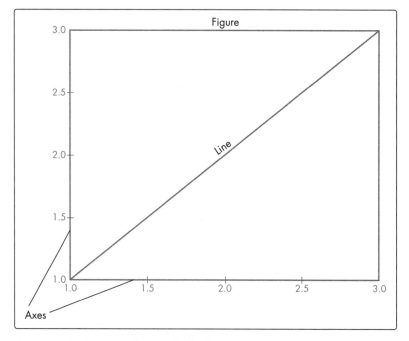

Figure 6-1: Architecture of a matplotlib plot

---

1. To learn more, see Chapter 11, "matplotlib," by John Hunter and Michael Droettboom in *The Architecture of Open Source Applications, Volume II: Structure, Scale, and a Few More Fearless Hacks* (2008; edited by Amy Brown and Greg Wilson; *http://www.aosabook.org/*).

The following program re-creates this plot, but we'll also explicitly create the Figure object and add axes to it, instead of just calling the plot() function and relying on it to create those:

```
>>> import matplotlib.pyplot as plt
>>> x = [1, 2, 3]
>>> y = [1, 2, 3]
❶ >>> fig = plt.figure()
❷ >>> ax = plt.axes()
>>> plt.plot(x, y)
[<matplotlib.lines.Line2D object at 0x7f9bad1dcc18>]
>>> plt.show()
>>>
```

Here, we create the Figure object using the figure() function at ❶, and then we create the axes using the axes() function at ❷. The axes() function also adds the axes to the Figure object. The last two lines are the same as in the earlier program. This time, when we call the plot() function, it sees that a Figure object with an Axes object already exists and directly proceeds to plot the data supplied to it.

Besides manually creating Figure and Axes objects, you can use two different functions in the pyplot module to get a reference to the current Figure and Axes objects. When you call the gcf() function, it returns a reference to the current Figure, and when you call the gca() function, it returns a reference to the current Axes. An interesting feature of these functions is that each will create the respective object if it doesn't already exist. How these functions work will become clearer as we make use of them later in this chapter.

Drawing a Circle

To draw a circle, you can add the Circle patch to the current Axes object, as demonstrated by the following example:

```
'''
Example of using matplotlib's Circle patch
'''
import matplotlib.pyplot as plt

def create_circle():
❶    circle = plt.Circle((0, 0), radius = 0.5)
    return circle

def show_shape(patch):
❷    ax = plt.gca()
    ax.add_patch(patch)
    plt.axis('scaled')
    plt.show()

if __name__ == '__main__':
❸    c = create_circle()
    show_shape(c)
```

In this program, we've separated the creation of the Circle patch object and the addition of the patch to the figure into two functions: create_circle() and show_shape(). In create_circle(), we make a circle with a center at (0, 0) and a radius of 0.5 by creating a Circle object with the coordinates of the center (0, 0) passed as a tuple and with the radius of 0.5 passed using the keyword argument of the same name at ❶. The function returns the created Circle object.

The show_shape() function is written such that it will work with any matplotlib patch. It first gets a reference to the current Axes object using the gca() function at ❷. Then, it adds the patch passed to it using the add_patch() function and, finally, calls the show() function to display the figure. We call the axis() function here with the scaled parameter, which basically tells matplotlib to automatically adjust the axis limits. We'll need to have this statement in all programs that use patches to automatically scale the axes. You can, of course, also specify fixed values for the limits, as we saw in Chapter 2.

At ❸, we call the create_circle() function using the label c to refer to the returned Circle object. Then, we call the show_shape() function, passing c as an argument. When you run the program, you'll see a matplotlib window showing the circle (see Figure 6-2).

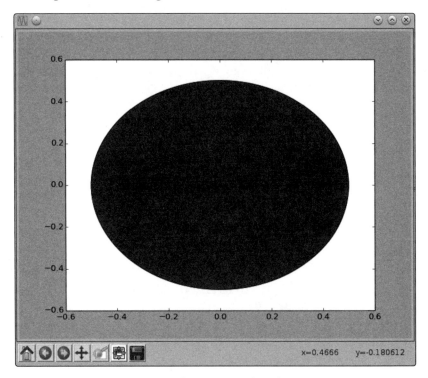

Figure 6-2: A circle with a center of (0, 0) and radius of 0.5

The circle doesn't quite look like a circle here, as you can see. This is due to the automatic aspect ratio, which determines the ratio of the length of the *x*- and *y*-axes. If you insert the statement ax.set_aspect('equal') after ❷, you will see that the circle does indeed look like a circle. The set_aspect() function is used to set the aspect ratio of the graph; using the equal argument, we ask matplotlib to set the ratio of the length of the *x*- and *y*-axes to 1:1.

Both the edge color and the face color (fill color) of the patch can be changed using the ec and fc keyword arguments. For example, passing fc='g' and ec='r' will create a circle with a green face color and red edge color.

Matplotlib supports a number of other patches, such as Ellipse, Polygon, and Rectangle.

Creating Animated Figures

Sometimes we may want to create figures with moving shapes. Matplotlib's animation support will help us achieve this. At the end of this section, we'll create an animated version of the projectile trajectory-drawing program.

First, let's see a simpler example. We'll draw a matplotlib figure with a circle that starts off small and grows to a certain radius indefinitely (unless the matplotlib window is closed):

```
'''
A growing circle
'''

from matplotlib import pyplot as plt
from matplotlib import animation

def create_circle():
    circle = plt.Circle((0, 0), 0.05)
    return circle

def update_radius(i, circle):
    circle.radius = i*0.5
    return circle,

def create_animation():
❶    fig = plt.gcf()
     ax = plt.axes(xlim=(-10, 10), ylim=(-10, 10))
     ax.set_aspect('equal')
     circle = create_circle()
❷    ax.add_patch(circle)
❸    anim = animation.FuncAnimation(
         fig, update_radius, fargs = (circle,), frames=30, interval=50)
     plt.title('Simple Circle Animation')
     plt.show()

if __name__ == '__main__':
    create_animation()
```

We start by importing the `animation` module from the matplotlib package. The `create_animation()` function carries out the core functionality here. It gets a reference to the current `Figure` object using the `gcf()` function at ❶ and then creates the axes with limits of –10 and 10 for both the x- and y-axes. After that, it creates a `Circle` object that represents a circle with a radius of 0.05 and a center at (0, 0) and adds this circle to the current axes at ❷. Then, we create a `FuncAnimation` object ❸, which passes the following data about the animation we want to create:

fig   This is the current `Figure` object.

update_radius   This function will be responsible for drawing *every* frame. It takes two arguments—a frame number that is automatically passed to it when called and the patch object that we want to update every frame. This function also must return the object.

fargs   This tuple consists of all the arguments to be passed to the `update_radius()` function other than the frame number. If there are no such arguments to pass, this keyword argument need not be specified.

frames   This is the number of frames in the animation. Our function `update_radius()` is called this many times. Here, we've arbitrarily chosen 30 frames.

interval   This is the time interval in milliseconds between two frames. If your animation seems too slow, decrease this value; if it seems too fast, increase this value.

We then set a title using the `title()` function and, finally, show the figure using the `show()` function.

As mentioned earlier, the `update_radius()` function is responsible for updating the property of the circle that will change each frame. Here, we set the radius to `i*0.5`, where `i` is the frame number. As a result, you see a circle that grows every frame for 30 frames—thus, the radius of the largest circle is 15. Because the axes' limits are set at –10 and 10, this gives the effect of the circle exceeding the figure's dimensions. When you run the program, you'll see your first animated figure, as shown in Figure 6-3.

You'll notice that the animation continues until you close the matplotlib window. This is the default behavior, which you can change by setting the keyword argument to `repeat=False` when you create the `FuncAnimation` object.

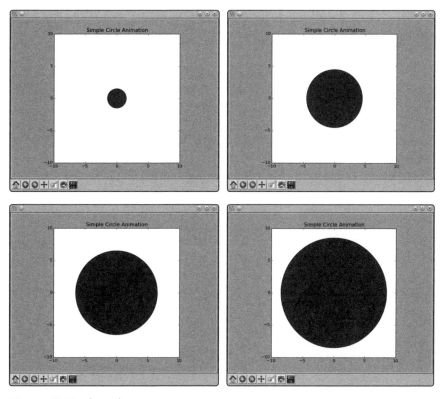

Figure 6-3: Simple circle animation

FUNCANIMATION OBJECT AND PERSISTENCE

You probably noted in the animated circle program that we assigned the created FuncAnimation object to the label anim even though we don't use it again elsewhere. This is because of an issue with matplotlib's current behavior—it doesn't store any reference to the FuncAnimation object, making it subject to garbage collection by Python. This means the animation will not be created. Creating a label referring to the object prevents this from happening.

For more on this issue, you may want to follow the discussions at *https://github.com/matplotlib/matplotlib/issues/1656/*.

Animating a Projectile's Trajectory

In Chapter 2, we drew the trajectory for a ball in projectile motion. Here, we'll build upon this drawing, making use of matplotlib's animation support to animate the trajectory so that it will come closer to demonstrating how you'd see a ball travel in real life:

```
'''
Animate the trajectory of an object in projectile motion
'''

from matplotlib import pyplot as plt
from matplotlib import animation
import math

g = 9.8

def get_intervals(u, theta):

    t_flight = 2*u*math.sin(theta)/g
    intervals = []
    start = 0
    interval = 0.005
    while start < t_flight:
        intervals.append(start)
        start = start + interval
    return intervals

def update_position(i, circle, intervals, u, theta):

    t = intervals[i]
    x = u*math.cos(theta)*t
    y = u*math.sin(theta)*t - 0.5*g*t*t
    circle.center = x, y
    return circle,

def create_animation(u, theta):

    intervals = get_intervals(u, theta)

    xmin = 0
    xmax = u*math.cos(theta)*intervals[-1]
    ymin = 0
    t_max = u*math.sin(theta)/g
    ymax = u*math.sin(theta)*t_max - 0.5*g*t_max**2
    fig = plt.gcf()
    ax = plt.axes(xlim=(xmin, xmax), ylim=(ymin, ymax))

    circle = plt.Circle((xmin, ymin), 1.0)
    ax.add_patch(circle)
```

❶ (marker at `ymax = u*math.sin(theta)*t_max - 0.5*g*t_max**2`)

❷ (marker at `ax = plt.axes(xlim=(xmin, xmax), ylim=(ymin, ymax))`)

```
❸      anim = animation.FuncAnimation(fig, update_position,
                            fargs=(circle, intervals, u, theta),
                            frames=len(intervals), interval=1,
                            repeat=False)

       plt.title('Projectile Motion')
       plt.xlabel('X')
       plt.ylabel('Y')
       plt.show()

if __name__ == '__main__':
    try:
        u = float(input('Enter the initial velocity (m/s): '))
        theta = float(input('Enter the angle of projection (degrees): '))
    except ValueError:
        print('You entered an invalid input')
    else:
        theta = math.radians(theta)
        create_animation(u, theta)
```

The create_animation() function accepts two arguments: u and theta. These arguments correspond to the initial velocity and the angle of projection (θ), which were supplied as input to the program. The get_intervals() function is used to find the time intervals at which to calculate the x- and y-coordinates. This function is implemented by making use of the same logic we used in Chapter 2, when we implemented a separate function, frange(), to help us.

To set up the axis limits for the animation, we'll need to find the minimum and maximum values of x and y. The minimum value for each is 0, which is the initial value for each. The maximum value of the x-coordinate is the value of the coordinate at the end of the flight of the ball, which is the last time interval in the list intervals. The maximum value of the y-coordinate is when the ball is at its highest point—that is, at ❶, where we calculate that point using the formula

$$t = \frac{u \sin \theta}{g}.$$

Once we have the values, we create the axes at ❷, passing the appropriate axis limits. In the next two statements, we create a representation of the ball and add it to the figure's Axes object by creating a circle of radius 1.0 at (xmin, ymin)—the minimum coordinates of the x- and y-axes, respectively.

We then create the FuncAnimation object ❸, supplying it with the current figure object and the following arguments:

update_position This function will change the center of the circle in each frame. The idea here is that a new frame is created for every time interval, so we set the number of frames to the size of the time intervals (see the description of frames in this list). We calculate the x- and y-coordinates of the ball at the time instant at the ith time interval, and we set the center of the circle to these values.

fargs The update_position() function needs to access the list of time intervals, intervals, initial velocity, and theta, which are specified using this keyword argument.

frames Because we'll draw one frame per time interval, we set the number of frames to the size of the intervals list.

repeat As we discussed in the first animation example, animation repeats indefinitely by default. We don't want that to happen in this case, so we set this keyword to False.

When you run the program, it asks for the initial inputs and then creates the animation, as shown in Figure 6-4.

Figure 6-4: Animation of the trajectory of a projectile

Drawing Fractals

Fractals are complex geometric patterns or shapes arising out of surprisingly simple mathematical formulas. Compared to geometric shapes, such as circles and rectangles, a fractal seems irregular and without any obvious pattern or description, but if you look closely, you see that patterns emerge and the entire shape is composed of numerous copies of itself. Because fractals involve the repetitive application of the same *geometric transformation* of points in a plane, computer programs are well-suited to create them. In this chapter, we'll learn how to draw the Barnsley fern, the Sierpiński triangle, and the Mandelbrot set (the latter two in the challenges)— popular examples of fractals studied in the field. Fractals abound in nature, too—popular examples include coastlines, trees, and snowflakes.

Transformations of Points in a Plane

A basic idea in creating fractals is that of the transformation of a point. Given a point, $P(x, y)$, in an x-y plane, an example of a transformation is $P(x, y) \rightarrow Q(x + 1, y + 1)$, which means that after applying the transformation, a new point, Q, which is one unit above and one unit to the right of P, is created. If you then consider Q as the starting point, you'll get another point, R, that's one unit above and one unit to the right of Q. Consider the starting point, P, to be $(1, 1)$. Figure 6-5 shows what the points would look like.

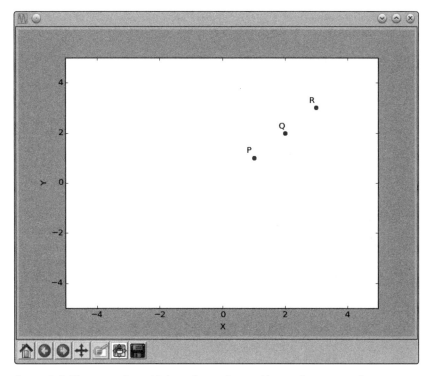

Figure 6-5: The points Q and R have been obtained by applying a transformation to the point P for two iterations.

This transformation is, thus, a rule describing how a point moves around in the x-y plane, starting from an initial position and moving to a different point at each iteration. We can think of a transformation as the point's *trajectory* in the plane. Now, consider that instead of one transformation rule, there are two such rules and one of these transformations is picked at *random* at every step. Let's consider these rules:

Rule 1: $P1\ (x, y) \rightarrow P2\ (x + 1, y - 1)$

Rule 2: $P1\ (x, y) \rightarrow P2\ (x + 1, y + 1)$

Consider $P1(1, 1)$ to be the starting point. If we carry out four iterations, we could have the following sequence of points:

$$P1\ (1, 1) \rightarrow P2\ (2, 0)\ \text{(Rule 1)}$$

$$P2\ (2, 0) \rightarrow P3\ (3, 1)\ \text{(Rule 2)}$$

$$P3\ (3, 1) \rightarrow P4\ (4, 2)\ \text{(Rule 2)}$$

$$P4\ (4, 2) \rightarrow P5\ (5, 1)\ \text{(Rule 1)}$$

. . . and so on.

The transformation rule is picked at random, with each rule having an equal probability of being selected. No matter which one is picked, the points will advance toward the right because we increase the x-coordinate in both cases. As the points go to the right, they move either up or down, thus creating a zigzag path. The following program charts out the path of a point when subjected to one of these transformations for a specified number of iterations:

```
'''
Example of selecting a transformation from two equally probable
transformations
'''
import matplotlib.pyplot as plt
import random

def transformation_1(p):
    x = p[0]
    y = p[1]
    return x + 1, y - 1

def transformation_2(p):
    x = p[0]
    y = p[1]
    return x + 1, y + 1

def transform(p):
    # List of transformation functions
❶    transformations = [transformation_1, transformation_2]
    # Pick a random transformation function and call it
❷    t = random.choice(transformations)
❸    x, y = t(p)
    return x, y

def build_trajectory(p, n):
    x = [p[0]]
    y = [p[1]]
    for i in range(n):
        p = transform(p)
        x.append(p[0])
        y.append(p[1])
```

```
    return x, y

if __name__ == '__main__':
    # Initial point
    p = (1, 1)
    n = int(input('Enter the number of iterations: '))
❹  x, y = build_trajectory(p, n)
    # Plot
❺  plt.plot(x, y)
    plt.xlabel('X')
    plt.ylabel('Y')
    plt.show()
```

We define two functions, transformation_1() and transformation_2(), corresponding to the two preceding transformations. In the transform() function, we create a list with these two function names at ❶ and use the random.choice() function to pick one of the transformations from the list at ❷. Now that we've picked the transformation to apply, we call it with the point, *P*, and store the coordinates of the transformed point in the labels x, y ❸ and return them.

SELECTING A RANDOM ELEMENT FROM A LIST

The random.choice() function we saw in our first fractal program can be used to select a random element from a list. Each element has an *equal* chance of being returned. Here's an example:

```
>>> import random
>>> l = [1, 2, 3]
>>> random.choice(l)
3
>>> random.choice(l)
1
>>> random.choice(l)
1
>>> random.choice(l)
3
>>> random.choice(l)
3
>>> random.choice(l)
2
```

The function also works with tuples and strings. In the latter case, it returns a random character from the string.

When you run the program, it asks you for the number of iterations, n—that is, the number of times the transformation would be applied. Then, it calls the build_trajectory() function with n and the initial point, P, which is set to (1, 1) ❹. The build_trajectory() function repeatedly calls the transform() function n times, using two lists, x and y, to store the x-coordinate and y-coordinate of all the transformed points. Finally, it returns the two lists, which are then plotted ❺.

Figures 6-6 and 6-7 show the trajectory of the point for 100 and 10,000 iterations, respectively. The zigzag motion is quite apparent in both figures. This zigzag path is usually referred to as a *random walk on a line.*

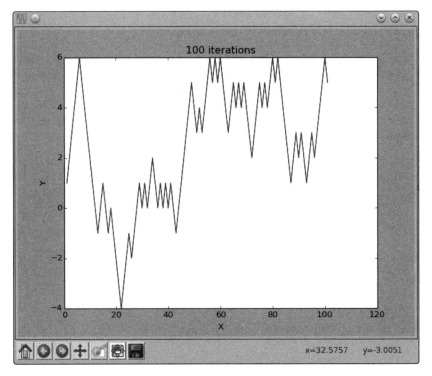

Figure 6-6: The zigzag path traced by the point (1, 1) when subjected to one or the other of the two transformations randomly for 100 iterations

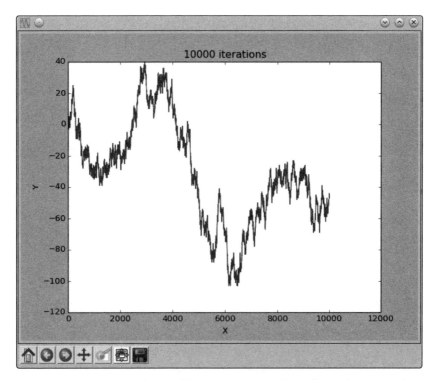

Figure 6-7: The zigzag path traced by the point (1, 1) when subjected to one or the other of the two transformations randomly for 10,000 iterations.

This example demonstrates a basic idea in creating fractals—starting from an initial point and applying a transformation to that point repeatedly. Next, we'll see an example of applying the same ideas to draw the *Barnsley fern*.

Drawing the Barnsley Fern

The British mathematician Michael Barnsley described how to create fern-like structures using repeated applications of a simple transformation on a point (see Figure 6-8).

Figure 6-8: Lady ferns[2]

He proposed the following steps to create fern-like structures: start with the point $(0, 0)$ and *randomly* select one of the following transformations with the assigned *probability*:

Transformation 1 (0.85 probability):

$$x_{n+1} = 0.85x_n + 0.04y_n$$

$$y_{n+1} = -0.04y_n + 0.85y_n + 1.6$$

Transformation 2 (0.07 probability):

$$x_{n+1} = 0.2x_n - 0.26y_n$$

$$y_{n+1} = 0.23y_n + 0.22y_n + 1.6$$

Transformation 3 (0.07 probability):

$$x_{n+1} = -0.15x_n - 0.28x_n$$

$$y_{n+1} = 0.26y_n + 0.24y_n + 0.44$$

Transformation 4 (0.01 probability):

$$x_{n+1} = 0$$

$$y_{n+1} = 0.16y_n$$

Each of these transformations is responsible for creating a part of the fern. The first transformation selected with the highest probability—and hence the maximum number of times—creates the stem and the bottom fronds of the fern. The second and third transformations create the bottom frond on the left and the right, respectively, and the fourth transformation creates the stem of the fern.

This is an example of nonuniform probabilistic selection, which we first learned about in Chapter 5. The following program draws the Barnsley fern for the specified number of points:

```
'''
Draw a Barnsley Fern
'''
import random
import matplotlib.pyplot as plt

def transformation_1(p):
    x = p[0]
    y = p[1]
    x1 = 0.85*x + 0.04*y
    y1 = -0.04*x + 0.85*y + 1.6
    return x1, y1

def transformation_2(p):
    x = p[0]
    y = p[1]
    x1 = 0.2*x - 0.26*y
    y1 = 0.23*x + 0.22*y + 1.6
    return x1, y1

def transformation_3(p):
    x = p[0]
    y = p[1]
    x1 = -0.15*x + 0.28*y
    y1 = 0.26*x  + 0.24*y + 0.44
    return x1, y1

def transformation_4(p):
    x = p[0]
    y = p[1]
    x1 = 0
    y1 = 0.16*y
    return x1, y1
```

```
def get_index(probability):
    r = random.random()
    c_probability = 0
    sum_probability = []
    for p in probability:
        c_probability += p
        sum_probability.append(c_probability)
    for item, sp in enumerate(sum_probability):
        if r <= sp:
            return item
    return len(probability)-1

def transform(p):
    # List of transformation functions
    transformations = [transformation_1, transformation_2,
                        transformation_3, transformation_4]
❶   probability = [0.85, 0.07, 0.07, 0.01]
    # Pick a random transformation function and call it
    tindex = get_index(probability)
❷   t = transformations[tindex]
    x, y = t(p)
    return x, y

def draw_fern(n):
    # We start with (0, 0)
    x = [0]
    y = [0]

    x1, y1 = 0, 0
    for i in range(n):
        x1, y1 = transform((x1, y1))
        x.append(x1)
        y.append(y1)
    return x, y

if __name__ == '__main__':
    n = int(input('Enter the number of points in the Fern: '))
    x, y = draw_fern(n)
    # Plot the points
    plt.plot(x, y, 'o')
    plt.title('Fern with {0} points'.format(n))
    plt.show()
```

When you run this program, it asks for the number of points in the fern to be specified and then creates the fern. Figures 6-9 and 6-10 show ferns with 1,000 and 10,000 points, respectively.

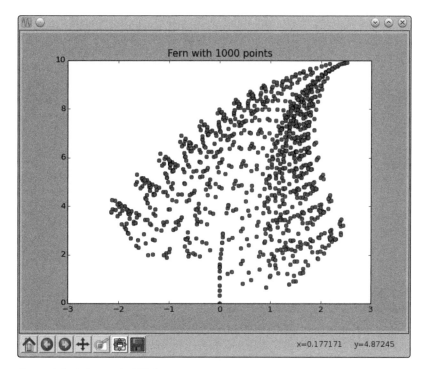

Figure 6-9: A fern with 1,000 points

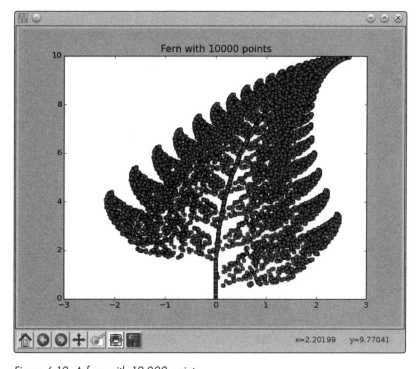

Figure 6-10: A fern with 10,000 points

The four transformation rules are defined in the `transformation_1()`, `transformation_2()`, `transformation_3()`, and `transformation_4()` functions. The probability of each being selected is declared in a list at ❶, and then one of them is selected ❷ to be applied every time the `transform()` function is called by the `draw_fern()` function.

The number of times the initial point (0, 0) is transformed is the same as the number of points in the fern specified as input to the program.

What You Learned

In this chapter, you started by learning how to draw basic geometric shapes and how to animate them. This process introduced you to a number of new matplotlib features. You then learned about geometric transformations and saw how repetitive simple transformations help you draw complex geometric shapes called *fractals*.

Programming Challenges

Here are a few programming challenges that should help you further apply what you've learned. You can find sample solutions at *http://www.nostarch .com/doingmathwithpython/*.

#1: Packing Circles into a Square

I mentioned earlier that matplotlib supports the creation of other geometric shapes. The `Polygon` patch is especially interesting, as it allows you to draw polygons with different numbers of sides. Here's how we can draw a square (each side of length 4):

```
'''
Draw a square
'''

from matplotlib import pyplot as plt

def draw_square():
    ax = plt.axes(xlim = (0, 6), ylim = (0, 6))
    square = plt.Polygon([(1, 1), (5, 1), (5, 5), (1, 5)], closed = True)
    ax.add_patch(square)
    plt.show()

if __name__ == '__main__':
    draw_square()
```

The `Polygon` object is created by passing the list of the vertices' coordinates as the first argument. Because we're drawing a square, we pass the coordinates of the four vertices: (1, 1), (5, 1), (5, 5), and (1, 5). Passing `closed=True` tells matplotlib that we want to draw a closed polygon, where the starting and the ending vertices are the same.

In this challenge, you'll attempt a very simplified version of the "circles packed into a square" problem. How many circles of radius 0.5 will fit in the square produced by this code? Draw and find out! Figure 6-11 shows how the final image will look.

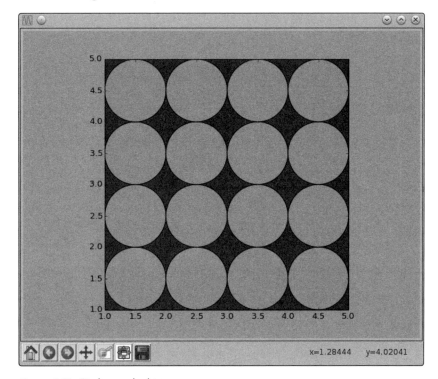

Figure 6-11: Circles packed into a square

The trick here is to start from the lower-left corner of the square—that is, (1, 1)—and then continue adding circles until the entire square is filled. The following snippet shows how you can create the circles and add them to the figure:

```
y = 1.5
while y < 5:
    x = 1.5
    while x < 5:
        c = draw_circle(x, y)
        ax.add_patch(c)

        x += 1.0
    y += 1.0
```

A point worth noting here is that this is *not* the most optimal or, for that matter, the only way to pack circles into a square, and finding different ways of solving this problem is popular among mathematics enthusiasts.

#2: Drawing the Sierpiński Triangle

The Sierpiński triangle, named after the Polish mathematician Wacław Sierpiński, is a fractal that is an equilateral triangle composed of smaller equilateral triangles embedded inside it. Figure 6-12 shows a Sierpiński triangle composed of 10,000 points.

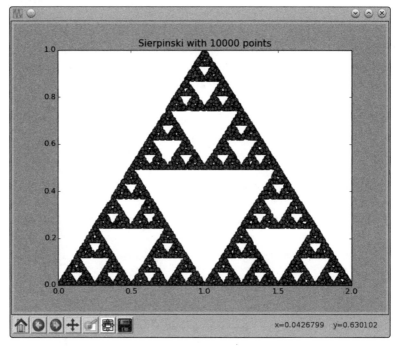

Figure 6-12: Sierpiński triangle with 10,000 points

The interesting thing here is that the same process that we used to draw a fern will also draw the Sierpiński triangle—only the transformation rules and their probability will change. Here's how you can draw the Sierpiński triangle: start with the point (0, 0) and apply one of the following transformations:

Transformation 1:

$$x_{n+1} = 0.5x_n$$

$$y_{n+1} = 0.5y_n$$

Transformation 2:

$$x_{n+1} = 0.5x_n + 0.5$$

$$y_{n+1} = 0.5y_n + 0.5$$

Transformation 3:

$$x_{n+1} = 0.5x_n + 1$$

$$y_{n+1} = 0.5y_n$$

Each of the transformations has an equal probability of being selected—1/3. Your challenge here is to write a program that draws the Sierpiński triangle composed of a certain number of points specified as input.

#3: Exploring Hénon's Function

In 1976, Michel Hénon introduced the Hénon function, which describes a transformation rule for a point $P(x, y)$ as follows:

$$P(x, y) \rightarrow Q(y + 1 - 1.4x^2, 0.3x)$$

Irrespective of the initial point (provided it's not very far from the origin), you'll see that as you create more points, they start lying along curved lines, as shown in Figure 6-13.

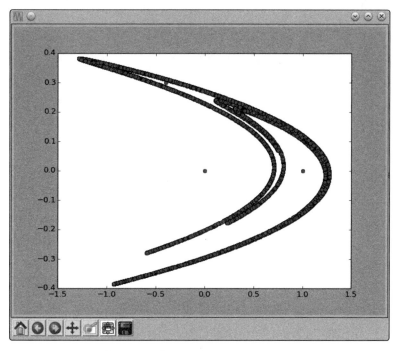

Figure 6-13: Hénon function with 10,000 points

Your challenge here is to write a program to create a graph showing 20,000 iterations of this transformation, starting with the point (1, 1).

Extra credit for writing another program to create an animated figure showing the points starting to lie along the curves! See *https://www.youtube.com/watch?v=76ll818RlpQ* for an example.

This is an example of a dynamical system, and the curved lines that all the points seem attracted to are referred to as *attractors*. To learn more about this function, dynamical systems, and fractals in general, you may want to refer to *Fractals: A Very Short Introduction* by Kenneth Falconer (Oxford University Press, 2013).

#4: Drawing the Mandelbrot Set

Your challenge here is to write a program to draw the *Mandelbrot set*—another example of the application of simple rules leading to a complicated-looking shape (see Figure 6-14). Before I lay down the steps to do that, however, we'll first learn about matplotlib's `imshow()` function.

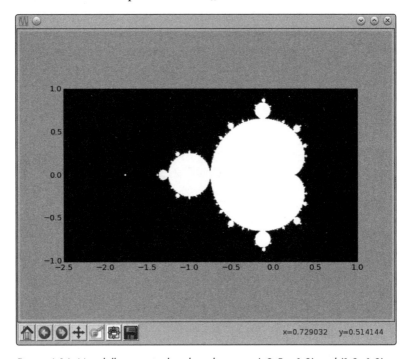

Figure 6-14: Mandelbrot set in the plane between (–2.5, –1.0) and (1.0, 1.0)

The imshow() Function

The `imshow()` function is usually used to display an external image, such as a JPEG or PNG image. You can see an example at *http://matplotlib.org/users/image_tutorial.html*. Here, however, we'll use the function to draw a new image of our own creation via matplotlib.

Consider the part of the Cartesian plane where x and y both range from 0 to 5. Now, consider six equidistant points along each axis: (0, 1, 2, 3, 4, 5) along the x-axis and the same set of points along the y-axis. If we take the Cartesian product of these points, we get 36 equally spaced points in the x-y plane with the coordinates (0, 0), (0, 1) . . . (0, 5), (1, 0), (1, 1) . . . (1, 5) . . . (5, 5). Let's now say that we want to color each of these points with a shade of gray—that is, some of these points will be black, some will be white, and others will be colored with a shade in between, randomly chosen. Figure 6-15 illustrates the scenario.

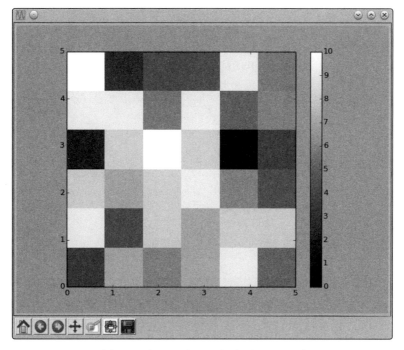

Figure 6-15: Part of the x-y plane with x and y both ranging from 0 to 5. We've considered 36 points in the region equidistant from each other and colored each with a shade of gray.

To create this figure, we have to make a list of six lists. Each of these six lists will in turn consist of six integers ranging from 0 to 10. Each number will correspond to the color for each point, 0 standing for black and 10 standing for white. We'll then pass this list to the imshow() function along with other necessary arguments.

Creating a List of Lists

A list can also contain lists as its members:

```
>>> l1 = [1, 2, 3]
>>> l2 = [4, 5, 6]
>>> l = [l1, l2]
```

Here, we created a list, l, consisting of two lists, l1 and l2. The first element of the list, l[0], is thus the same as the l1 list and the second element of the list, l[1], is the same as the l2 list:

```
>>> l[0]
[1, 2, 3]
```

```
>>> l[1]
[4, 5, 6]
```

To refer to an individual element within one of the member lists, we have to specify two indices—l[0][1] refers to the second element of the first list, l[1][2] refers to the third element of the second list, and so on.

Now that we know how to work with a list of lists, we can write the program to create a figure similar to Figure 6-15:

```
import matplotlib.pyplot as plt
import matplotlib.cm as cm
import random

❶ def initialize_image(x_p, y_p):
      image = []
      for i in range(y_p):
          x_colors = []
          for j in range(x_p):
              x_colors.append(0)
          image.append(x_colors)
      return image

  def color_points():
      x_p = 6
      y_p = 6
      image = initialize_image(x_p, y_p)
      for i in range(y_p):
          for j in range(x_p):
❷             image[i][j] = random.randint(0, 10)
❸     plt.imshow(image, origin='lower', extent=(0, 5, 0, 5),
              cmap=cm.Greys_r, interpolation='nearest')
      plt.colorbar()
      plt.show()

  if __name__ == '__main__':
      color_points()
```

The initialize_image() function at ❶ creates a list of lists with each of the elements initialized to 0. It accepts two arguments, x_p and y_p, which correspond to the number of points along the *x*-axis and *y*-axis, respectively. This effectively means that the initialized list image will consist of x_p lists with each list containing y_p zeros.

In the color_points() function, once you have the image list back from initialize_image(), assign a random integer between 0 and 10 to the element image[i][j] at ❷. When we assign this random integer to the element, we are assigning a color to the point in the Cartesian plane that's *i* steps along the *y*-axis and *j* steps along the *x*-axis from the origin. It's important to note that the imshow() function automatically deduces the color of a point from its position in the image list and doesn't care about its specific *x*- and *y*-coordinates.

Then, call the `imshow()` function at ❸, passing `image` as the first argument. The keyword argument `origin='lower'` specifies that the number in `image[0][0]` corresponds to the color of the point $(0, 0)$. The keyword argument `extent=(0, 5, 0, 5)` sets the lower-left and upper-right corners of the image to $(0, 0)$ and $(5, 5)$, respectively. The keyword argument `cmap=cm.Greys_r` specifies that we're going to create a grayscale image.

The last keyword argument, `interpolation='nearest'`, specifies that matplotlib should color a point for which the color wasn't specified with the same color as the one nearest to it. What does this mean? Note that we consider and specify the color for only 36 points in the region $(0, 5)$ and $(5, 5)$. Because there is an infinite number of points in this region, we tell matplotlib to set the color of an unspecified point to that of its nearest point. This is the reason you see color "boxes" around each point in the figure.

Call the `colorbar()` function to display a color bar in the figure showing which integer corresponds to which color. Finally, call `show()` to display the image. Note that due to the use of the `random.randint()` function, your image will be colored differently than the one in Figure 6-15.

If you increase the number of points along each axis by setting `x_p` and `y_p` to, let's say, 20 in `color_points()`, you'll see a figure similar to the one shown in Figure 6-16. Note that the color boxes grow smaller in size. If you increase the number of points even more, you'll see the size of the boxes shrink further, giving the illusion that each point has a different color.

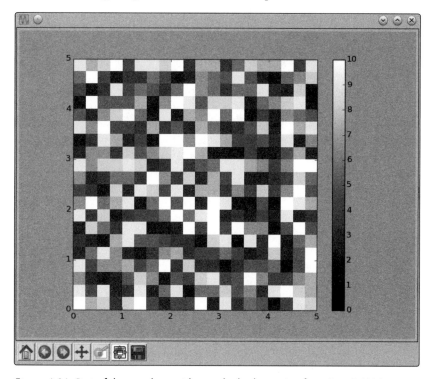

Figure 6-16: Part of the x-y plane with x and y both ranging from 0 to 5. We've considered 400 points in the region equidistant from each other and colored each with a shade of gray.

Drawing the Mandelbrot Set

We'll consider the area of the *x-y* plane between (–2.5, –1.0) and (1.0, 1.0) and divide each axis into 400 equally spaced points. The Cartesian product of these points will give us 1,600 equally spaced points in this region. We'll refer to these points as $(x_1, y_1), (x_1, y_2) \ldots (x_{400}, y_{400})$.

Create a list, image, by calling the initialize_image() function we saw earlier with both x_p and y_p set to 400. Then, follow these steps for *each* of the generated points (x_i, y_k):

1. First, create two complex numbers, $z_1 = 0 + 0j$ and $c = x_i + y_k j$. (Recall that we use j for $\sqrt{-1}$.)

2. Create a label iteration and set it to 0—that is, iteration=0.

3. Create a complex number, $z_1 = z_1^2 + c$.

4. Increment the value stored in iteration by 1—that is, iteration = iteration + 1.

5. If abs(z1) < 2 and iteration < max_iteration, then go back to step 3; otherwise, go to step 6. The larger the value of max_iteration, the more detailed the image, but the longer it'll take to create the image. Set max_iteration to 1,000 here.

6. Set the color of the point (x_i, y_k) to the value in iteration—that is, image[k][i] = iteration.

Once you have the complete image list, call the imshow() function with the extent keyword argument changed to indicate the region bounded by (–2.5, –1.0) and (1.0, 1.0).

This algorithm is usually referred to as the *escape-time algorithm*. When the maximum number of iterations is reached before a point's magnitude exceeds 2, that point belongs to the Mandelbrot set and is colored white. The points that exceed the magnitude within fewer iterations are said to "escape"; they don't belong to the Mandelbrot set and are colored black. You can experiment by decreasing and increasing the number of points along each axis. Decreasing the number of points will lead to a grainy image, while increasing them will result in a more detailed image.

7

SOLVING CALCULUS PROBLEMS

In this final chapter, we'll learn to solve calculus problems. We'll first learn about mathematical functions, followed by a quick overview of the common mathematical functions available in Python's standard library and SymPy. Then, we'll learn how we can find the limits of functions and calculate derivatives and integrals—that is, the kinds of things you'd be doing in any calculus class. Let's get started!

What Is a Function?

Let's start out with some basic definitions. A function is a *mapping* between an input set and an output set. The special condition of a function is that an element of the input set is related to *exactly one* element of the output set. For example, Figure 7-1 shows two sets such that an element of the output set is the square of an element that belongs to the input set.

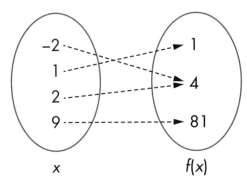

Figure 7-1: A function describes a mapping between an input set and an output set. Here, an element of the output set is the square of an element from the input set.

Using the familiar function notation, we'd write this function as $f(x) = x^2$, where x is the independent variable quantity. So $f(2) = 4$, $f(100) = 10000$, and so on. We refer to x as the independent variable quantity because we're free to assume a value for it as long as that value is within its domain (see the next section).

Functions can also be defined in terms of multiple variables. For example, $f(x, y) = x^2 + y^2$ defines a function of two variables, x and y.

Domain and Range of a Function

The *domain* of a function is the set of input values that the independent variable can validly assume. The output set of a function is called the *range*.

For example, the domain of the function $f(x) = 1/x$ is all nonzero real and complex numbers because $1/0$ isn't defined. The range is formed by the set of values obtained by substituting each number in the domain into $1/x$, so in this case it is also all nonzero real and complex numbers.

NOTE *The domain and range of a function can certainly be different. For example, for the function x^2, the domain is all positive and negative numbers, but the range is only the positive numbers.*

An Overview of Common Mathematical Functions

We've already used a number of common mathematical functions from the Python standard library's math module. A couple of familiar examples are the sin() and cos() functions, which correspond to the trigonometric functions sine and cosine. Other trigonometric functions—`tan()` and the inverse equivalents of these functions, `asin()`, `acos()`, and `atan()`—are also defined.

The `math` module also includes functions that find the logarithm of a number—the natural logarithm function `log()`, the base-2 logarithm `log2()`, and the base-10 logarithm `log10()`—as well as the function `exp()`, which finds the value of e^x, where e is Euler's number (approximately 2.71828).

One drawback of all these functions is that they're not suitable for working with symbolic expressions. If we want to manipulate a mathematical expression involving symbols, we have to start using the equivalent functions defined by SymPy.

Let's see a quick example:

```
>>> import math
>>> math.sin(math.pi/2)
1.0
```

Here, we find the sine of the angle $\pi/2$ using the `sin()` function defined by the standard library's `math` module. Then, we can do the same using SymPy.

```
>>> import sympy
>>> sympy.sin(math.pi/2)
1.00000000000000
```

Similar to the standard library's `sin()` function, SymPy's `sin()` function expects the angle to be expressed in radians. Both functions return 1.

Now, let's try to call each function with a symbol instead and see what happens:

```
>>> from sympy import Symbol
>>> theta = Symbol('theta')
❶ >>> math.sin(theta) + math.sin(theta)
Traceback (most recent call last):
  File "<pyshell#53>", line 1, in <module>
    math.sin(theta) + math.sin(theta)
  File "/usr/lib/python3.4/site-packages/sympy/core/expr.py", line 225, in
__float__
    raise TypeError("can't convert expression to float")
TypeError: can't convert expression to float

❷ >>> sympy.sin(theta) + sympy.sin(theta)
2*sin(theta)
```

The standard library's `sin()` function doesn't know what to do when we call it with theta at ❶, so it raises an exception to indicate that it's expecting a numerical value as an argument to the `sin()` function. On the other hand, SymPy is able to perform the same operation at ❷, and it returns the expression 2*sin(theta) as the result. This is hardly surprising to us now, but it illustrates the kinds of tasks where the standard library's mathematical functions can fall short.

Let's consider another example. Say we want to derive the expression for the time it takes for a body in projectile motion to reach the highest point if it's thrown with initial velocity u at an angle theta (see "Projectile Motion" on page 48).

At the highest point, u*sin(theta)-g*t = 0, so to find t, we'll use the solve() function we learned about in Chapter 4:

```
>>> from sympy import sin, solve, Symbol
>>> u = Symbol('u')
>>> t = Symbol('t')
>>> g = Symbol('g')
>>> theta = Symbol('theta')
>>> solve(u*sin(theta)-g*t, t)
[u*sin(theta)/g]
```

The expression for t, as we learned earlier, turns out to be u*sin(theta)/g, and it illustrates how the solve() function can be used to find solutions to equations containing mathematical functions as well.

Assumptions in SymPy

In all our programs, we've created a Symbol object in SymPy, defining the variable like so: x = Symbol('x'). Assume that as a result of an operation you asked SymPy to perform, SymPy needs to check whether the expression $x + 5$ is greater than 0. Let's see what would happen:

```
>>> from sympy import Symbol
>>> x = Symbol('x')
>>> if (x+5) > 0:
    print('Do Something')
else:
    print('Do Something else')

Traceback (most recent call last):
  File "<pyshell#45>", line 1, in <module>
    if (x + 5) > 0:
  File "/usr/lib/python3.4/site-packages/sympy/core/relational.py", line 103,
in __nonzero__
    raise TypeError("cannot determine truth value of\n%s" % self)
TypeError: cannot determine truth value of
x + 5 > 0
```

Because SymPy doesn't know anything about the sign of x, it can't deduce whether $x + 5$ is greater than 0, so it displays an error. But basic math tells us that if x is positive, $x + 5$ will always be positive, and if x is negative, it will be positive only in certain cases.

So if we create a `Symbol` object specifying `positive=True`, we tell SymPy to assume only positive values. Now it knows for sure that $x + 5$ is definitely greater than 0:

```
>>> x = Symbol('x', positive=True)
>>> if (x+5) > 0:
    print('Do Something')
else:
    print('Do Something else')

Do Something
```

Note that if we'd instead specified `negative=True`, we could get the same error as in the first case. Just as we can declare a symbol as `positive` and `negative`, it's also possible to specify it as `real`, `integer`, `complex`, `imaginary`, and so on. These declarations are referred to as *assumptions* in SymPy.

Finding the Limit of Functions

A common task in calculus is finding the *limiting value* (or simply the *limit*) of the function, when the variable's value is assumed to approach a certain value. Consider a function $f(x) = 1/x$, whose graph is shown in Figure 7-2.

As the value of x increases, the value of $f(x)$ approaches 0. Using the limit notation, we'd write this as

$$\lim_{x \to \infty} \frac{1}{x} = 0.$$

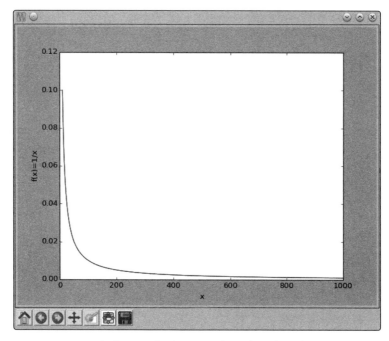

Figure 7-2: A graph showing the function 1/x as the value of x increases

We can find limits of functions in SymPy by creating objects of the Limit class as follows:

```
❶ >>> from sympy import Limit, Symbol, S
❷ >>> x = Symbol('x')
❸ >>> Limit(1/x, x, S.Infinity)
   Limit(1/x, x, oo, dir='-')
```

At ❶, we import the Limit and Symbol classes, as well as S, which is a special SymPy class that contains the definition of infinity (positive and negative) and other special values. Then at ❷ we create a symbol object, x, to represent *x*. We create the Limit object at ❸, passing it three arguments: 1/x, the variable x, and finally the value at which we want to calculate the function's limit (infinity, given by S.Infinity).

The result is returned as an *unevaluated* object with the oo symbol denoting positive infinity and the dir='-' symbol specifying that we are approaching the limit from the negative side.

To find the value of the limit, we use the doit() method:

```
>>> l = Limit(1/x, x, S.Infinity)
>>> l.doit()
0
```

By default, the limit is found from a positive direction, unless the value at which the limit is to be calculated is positive or negative infinity. In the case of positive infinity, the direction is negative, and vice versa. You can change the default direction as follows:

```
>>> Limit(1/x, x, 0, dir='-').doit()
-oo
```

Here, we calculate

$$\lim_{x \to 0} \frac{1}{x},$$

and as we approach 0 for *x* from the negative side, the value of the limit approaches negative infinity. On the other hand, if we approach 0 from the positive side, the value approaches positive infinity:

```
>>> Limit(1/x, x, 0, dir='+').doit()
oo
```

The Limit class also handles functions with limits of indeterminate forms,

$$\left(\frac{0}{0}, \frac{\inf}{\inf} \right),$$

automatically:

```
>>> from sympy import Symbol, sin
>>> Limit(sin(x)/x, x, 0).doit()
1
```

You have very likely used l'Hôpital's rule to find such limits, but as we see here, the Limit class takes care of this for us.

Continuous Compound Interest

Say you've deposited $1 in a bank. This deposit is the *principal*, which pays you *interest*—in this case, interest of 100 percent that compounds n times yearly for 1 year. The amount you'll get at the end of 1 year is given by

$$A = \left(1 + \frac{1}{n}\right)^n.$$

The prominent mathematician James Bernoulli discovered that as the value of n increases, the term $(1 + 1/n)^n$ approaches the value of e—the constant that we can verify by finding the limit of the function:

```
>>> from sympy import Limit, Symbol, S
>>> n = Symbol('n')
>>> Limit((1+1/n)**n, n, S.Infinity).doit()
E
```

For any principal amount p, any rate r, and any number of years t, the compound interest is calculated using the formula

$$A = P\left(1 + \frac{r}{n}\right)^{nt}.$$

Assuming continuous compounding interest, we can find the expression for A as follows:

```
>>> from sympy import Symbol, Limit, S
>>> p = Symbol('p', positive=True)
>>> r = Symbol('r', positive=True)
>>> t = Symbol('t', positive=True)
>>> Limit(p*(1+r/n)**(n*t), n, S.Infinity).doit()
p*exp(r*t)
```

We create three symbol objects, representing the principal amount, p, the rate of interest, r, and the number of years, t. We also tell SymPy that these symbols will assume positive values by passing the positive=True keyword argument while creating the Symbol objects. If we don't specify, SymPy doesn't know anything about the numerical values the symbol can assume and may not be able to evaluate the limit correctly. We then feed in the expression for the compound interest to create the Limit object and evaluate it using the `doit()` method. The limit turns out to be `p*exp(r*t)`, which tells us that the compound interest grows exponentially with time for the fixed rate of interest.

Instantaneous Rate of Change

Consider a car moving along a road. It accelerates uniformly such that the distance traveled, S, is given by the function

$$S(t) = 5t^2 + 2t + 8.$$

In this function, the independent variable is t, which represents the time elapsed since the car started moving.

If we measure the distance traveled in time t_1 and time t_2 such that $t_2 > t_1$, then we can calculate the distance moved by the car in 1 unit of time using the expression

$$\frac{S(t_2) - S(t_1)}{t_2 - t_1}.$$

This is also referred to as the average rate of change of the function $S(t)$ with respect to the variable t, or in other words, the average speed. If we write t_2 as $t_1 + \delta_t$—where δ_t is the difference between t_2 and t_1 in units of time—we can rewrite the expression for the average speed as

$$\frac{S(t_1 + \delta_t) - S(t_1)}{\delta_t}.$$

This expression is also a function with t_1 as the variable. Now, if we further assume δ_t to be really small, such that it approaches 0, we can use limit notation to write this as

$$\lim_{\delta_t \to 0} \frac{S(t_1 + \delta_t) - S(t_1)}{\delta_t}.$$

We will now evaluate the above limit. First, let's create the various expression objects:

```
>>> from sympy import Symbol, Limit
>>> t = Symbol('t')
❶ >>> St = 5*t**2 + 2*t + 8

>>> t1 = Symbol('t1')
>>> delta_t = Symbol('delta_t')

❷ >>> St1 = St.subs({t: t1})
❸ >>> St1_delta = St.subs({t: t1 + delta_t})
```

We first define the function $S(t)$ at ❶. Then, we define two symbols, t1 and delta_t, which correspond to t_1 and δ_t. Using the `subs()` method, we then find $S(t_1)$ and $S(t_1 + \delta_t)$ by substituting in the value of t for t1 and t1_delta_t at ❷ and ❸, respectively.

Now, let's evaluate the limit:

```
>>> Limit((St1_delta-St1)/delta_t, delta_t, 0).doit()
10*t1 + 2
```

The limit turns out to be 10*t1 + 2, and it's the rate of change of $S(t)$ at time t1, or the instantaneous rate of change. This change is more commonly referred to as the *instantaneous speed* of the car at the time instant t1.

The limit we calculated here is referred to as the *derivative* of a function, and we can calculate it directly using SymPy's Derivative class.

Finding the Derivative of Functions

The derivative of a function $y = f(x)$ expresses the rate of change in the dependent variable, y, with respect to the independent variable, x. It's denoted as either $f'(x)$ or dy/dx. We can find the derivative of a function by creating an object of the Derivative class. Let's use the previous function representing the motion of a car as an example:

❶ ```
>>> from sympy import Symbol, Derivative

>>> t = Symbol('t')
>>> St = 5*t**2 + 2*t + 8
```

❷ ```
>>> Derivative(St, t)
Derivative(5*t**2 + 2*t + 8, t)
```

We import the Derivative class at ❶. At ❷, we create an object of the Derivative class. The two arguments passed while creating the object are the function St and the symbol t, which corresponds to the variable t. As with the Limit class, an object of the Derivative class is returned, and the derivative is not actually calculated. We call the doit() method on the unevaluated Derivative object to find the derivative:

```
>>> d = Derivative(St, t)
>>> d.doit()
10*t + 2
```

The expression for the derivative turns out to be 10*t + 2. Now, if we want to calculate the value of the derivative at a particular value of t—say, $t = t_1$ or $t = 1$—we can use the subs() method:

```
>>> d.doit().subs({t:t1})
10*t1 + 2
>>> d.doit().subs({t:1})
12
```

Let's try a complicated arbitrary function with x as the only variable: $(x^3 + x^2 + x) \times (x^2 + x)$.

```
>>> from sympy import Derivative, Symbol
>>> x = Symbol('x')
>>> f = (x**3 + x**2 + x)*(x**2+x)
>>> Derivative(f, x).doit()
(2*x + 1)*(x**3 + x**2 + x) + (x**2 + x)*(3*x**2 + 2*x + 1)
```

You may consider this function the product of two independent functions, which means that, by hand, we'd need to make use of the *product rule* of differentiation to find the derivative. But we don't need to worry about that here because we can just create an object of the `Derivative` class to do that for us.

Try out some other complicated expressions, such as expressions involving trigonometric functions.

A Derivative Calculator

Now let's write a derivative calculator program, which will take a function as input and then print the result of differentiating it with respect to the variable specified:

```
'''
Derivative calculator
'''

from sympy import Symbol, Derivative, sympify, pprint
from sympy.core.sympify import SympifyError

def derivative(f, var):
    var = Symbol(var)
    d = Derivative(f, var).doit()
    pprint(d)

if __name__=='__main__':
❶    f = input('Enter a function: ')
    var = input('Enter the variable to differentiate with respect to: ')
    try:
❷        f = sympify(f)
    except SympifyError:
        print('Invalid input')
    else:
❸        derivative(f, var)
```

At ❶, we ask the user to input a function for which the derivative is to be found, and then we ask for the variable with respect to which the function is to be differentiated. At ❷, we convert the input function into a SymPy object using the `sympify()` function. We call this function in a try...except block so that we can display an error message in case the user enters an invalid input. If the input expression is a valid expression, we call the derivative function at ❸, passing the converted expression and the variable with respect to which the function is to be differentiated as arguments.

In the derivative() function, we first create a Symbol object that corresponds to the variable with respect to which the function is to be differentiated. We use the label var to refer to this variable. Next, we create a Derivative object that passes both the function to differentiate and the symbol object var. We immediately call the doit() method to evaluate the derivative, and we then use the pprint() function to print the result so that it appears close to its mathematical counterpart. A sample execution of the program follows:

```
Enter a function: 2*x**2 + 3*x + 1
Enter the variable to differentiate with respect to: x
4·x + 3
```

Here's a sample run when used with a function of two variables:

```
Enter a function: 2*x**2 + y**2
Enter the variable to differentiate with respect to: x
4·x
```

Calculating Partial Derivatives

In the previous program, we saw that it's possible to calculate the derivative of a multivariable function with respect to any variable using the Derivative class. This calculation is usually referred to as *partial differentiation*, with *partial* indicating that we assume only one variable varies while the others are fixed.

Let's consider the function $f(x, y) = 2xy + xy^2$. The partial differentiation of $f(x, y)$ with respect to x is

$$\frac{\partial f}{\partial x} = 2y + y^2.$$

The preceding program is capable of finding the partial derivative because it's just a matter of specifying the right variable:

```
Enter a function: 2*x*y + x*y**2
Enter the variable to differentiate with respect to: x
y² + 2·y
```

NOTE *A key assumption I've made in this chapter is that all the functions we're calculating the derivative of are differentiable in their respective domains.*

Higher-Order Derivatives and Finding the Maxima and Minima

By default, creating the derivative object using the `Derivative` class finds the first-order derivative. To find higher-order derivatives, simply specify the order of the derivative to calculate as the third argument when you create the `Derivative` object. In this section I will show you how to use the first- and second-order derivative of the function to find its maxima and minima on an interval.

Consider the function $x^5 - 30x^3 + 50x$, defined on the domain $[-5, 5]$. Note that I have used square brackets to indicate a closed domain, indicating that the variable x can assume any real value greater than or equal to -5 and less than or equal to 5 (see Figure 7-3).

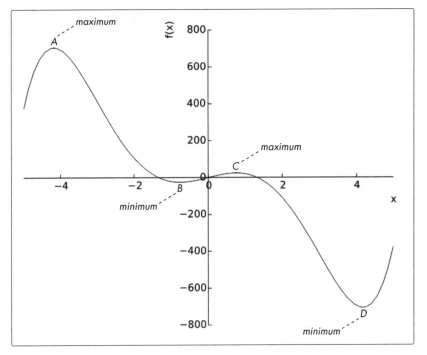

Figure 7-3: Plot of the function $x^5 - 30x^3 + 50x$, where $-5 \leq x \leq 5$

From the graph, we can see that the function attains its minimum value on the interval $-2 \leq x \leq 0$ at the point B. Similarly, it attains its maximum value on the interval $0 \leq x \leq 2$ at the point C. On the other hand, the function attains its maximum and minimum values on the entire domain of x that we've considered here at the points A and D, respectively. Thus, when we consider the function on the whole interval $[-5, 5]$, the points B and C are referred to as a *local minimum* and a *local maximum*, respectively, while the points A and D are the *global maximum* and the *global minimum*, respectively.

The term *extremum* (plural *extrema*) refers to the points where the function attains a local or global maximum or minimum. If x is an extremum of the function $f(x)$, then the first-order derivative of f at x, denoted $f'(x)$, must vanish. This property shows that a good way to find possible extrema is to try to solve the equation $f'(x) = 0$. Such solutions are called *critical points* of the function. Let's try this out:

```
>>> from sympy import Symbol, solve, Derivative
>>> x = Symbol('x')
>>> f = x**5 - 30*x**3 + 50*x
>>> d1 = Derivative(f, x).doit()
```

Now that we have calculated the first-order derivative, $f'(x)$, we'll solve $f'(x) = 0$ to find the critical points:

```
>>> critical_points = solve(d1)
>>> critical_points
[-sqrt(-sqrt(71) + 9), sqrt(-sqrt(71) + 9), -sqrt(sqrt(71) + 9),
sqrt(sqrt(71) + 9)]
```

The numbers in the list `critical_points` shown here correspond to the points *B*, *C*, *A*, and *D*, respectively. We will create labels to refer to these points, and then we can use the labels in our commands:

```
>>> A = critical_points[2]
>>> B = critical_points[0]
>>> C = critical_points[1]
>>> D = critical_points[3]
```

Because all the critical points for this function lie within the considered interval, they are all relevant for our search for the global maximum and minimum of $f(x)$. We may now apply the so-called *second derivative test* to narrow down which critical points could be global maxima or minima.

First, we calculate the second-order derivative for the function $f(x)$. Note that to do so, we enter 2 as the third argument:

```
>>> d2 = Derivative(f, x, 2).doit()
```

Now, we find the value of the second derivative by substituting the value of each of the critical points one by one in place of x. If the resulting value is less than 0, the point is a local maximum; if the value is greater than 0, it's a local minimum. If the resulting value is 0, then the test is inconclusive and we cannot deduce anything about whether the critical point x is a local minimum, maximum, or neither.

```
>>> d2.subs({x:B}).evalf()
127.661060789073
>>> d2.subs({x:C}).evalf()
-127.661060789073
>>> d2.subs({x:A}).evalf()
-703.493179468151
>>> d2.subs({x:D}).evalf()
703.493179468151
```

Evaluating the second derivative test at the critical points tells us that the points *A* and *C* are local maxima and the points *B* and *D* are local minima.

The global maximum and minimum of $f(x)$ on the interval $[-5, 5]$ is attained either at a critical point x or at one of the endpoints of the domain ($x = -5$ and $x = 5$). We have already found all of the critical points, which are the points *A*, *B*, *C*, and *D*. The function cannot attain its global minimum at either of the critical points *A* or *C* because they are local maximums. By similar logic, the function cannot attain its global maximum at *B* or *D*.

Thus, to find the global maximum, we must compute the value of $f(x)$ at the points *A*, *C*, −5, and 5. Among these points, the place where $f(x)$ has the largest value must be the global maximum.

We will create two labels, x_min and x_max, to refer to the domain boundaries and evaluate the function at the points A, C, x_min, and x_max:

```
>>> x_min = -5
>>> x_max = 5

>>> f.subs({x:A}).evalf()
705.959460380365
>>> f.subs({x:C}).evalf()
25.0846626340294
>>> f.subs({x:x_min}).evalf()
375.000000000000
>>> f.subs({x:x_max}).evalf()
-375.000000000000
```

By these calculations, as well as by examining the function value at all the critical points and the domain boundaries (Figure 7-3), we see that the point *A* turns out be the global maximum.

Similarly, to determine the global minimum, we must compute the values of $f(x)$ at the points *B*, *D*, −5, and 5:

```
>>> f.subs({x:B}).evalf()
-25.0846626340294
>>> f.subs({x:D}).evalf()
-705.959460380365
>>> f.subs({x:x_min}).evalf()
375.000000000000
>>> f.subs({x:x_max}).evalf()
-375.000000000000
```

The point where $f(x)$ has the smallest value must be the global minimum for the function; this turns out to be point *D*.

This method for finding the extrema of a function—by considering the function's value at all of the critical points (after potentially discarding some via the second derivative test) and boundary values—will always work as long as the function is twice differentiable. That is, both the first and second derivative must exist everywhere in the domain.

For a function such as e^x, there might not be any critical points in the domain, but in this case the method works fine: it simply tells us that the extrema occur at the domain boundary.

Finding the Global Maximum Using Gradient Ascent

Sometimes we're just interested in finding the global maximum for a function instead of all the local and global maxima and minima. For example, we might want to discover the angle of projection for which a ball will cover the maximum horizontal distance. We're going to learn a new, more practical approach to solve such a problem. This approach makes use of the first derivative only, so it's applicable only to functions for which the first derivative can be calculated.

This method is called the *gradient ascent method*, which is an iterative approach to finding the global maximum. Because the gradient ascent method involves lots of computation, it's the perfect kind of thing to solve programmatically rather than by hand. Let's try it out using the example problem of finding the angle of projection. In Chapter 2, we derived the expression

$$t_{\text{flight}} = 2\frac{u\sin\theta}{g}$$

to calculate the time of flight for a body in projectile motion that's thrown with a velocity u at an angle θ. The *range* of a projectile, R, is the total horizontal distance traveled by the projectile and is given by the product of $u_x \times t_{\text{flight}}$. Here, u_x is the horizontal component of the initial velocity and is equal to $u\cos\theta$. Substituting the formulas for u_x and t_{flight}, we get the expression

$$R = u\cos\theta \times \frac{2u\sin\theta}{g} = \frac{u^2\sin 2\theta}{g}.$$

The plot in Figure 7-4 shows values of θ between 0 and 90 degrees and the corresponding range (distance traveled) for each angle. From the graph, we can see that the maximum range is obtained when the angle of projection is around 45 degrees. We'll now learn to use the gradient ascent method to find this value of θ numerically.

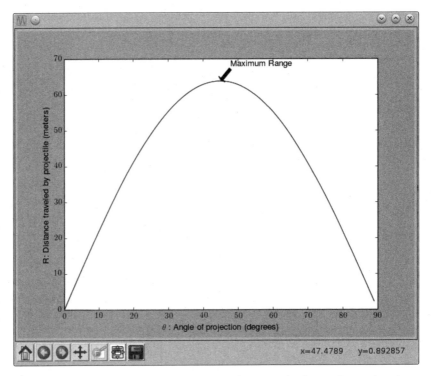

Figure 7-4: The range of a projectile thrown with an initial velocity of 25 m/s with varying angles of projection

The gradient ascent method is an iterative method: we start with an initial value of θ—say, 0.001, or $\theta_{old} = 0.001$—and gradually get closer to the value of θ that corresponds to the maximum range (Figure 7-5). The step that gets us closer is the equation

$$\theta_{new} = \theta_{old} + \lambda \frac{dR}{d\theta},$$

where λ is the *step size* and

$$\frac{dR}{d\theta}$$

is the derivative of R with respect to θ. Once we set $\theta_{old} = 0.001$, we do the following:

1. Calculate θ_{new} using the preceding equation.
2. If the absolute difference $\theta_{new} - \theta_{old}$ is greater than a value, ε, we set $\theta_{old} = \theta_{new}$ and return to step 1. Otherwise, we go to step 3.
3. θ_{new} is an approximate value of θ for which R has the maximum value.

The value of *epsilon* (ε) determines when we decide to stop the iteration of the algorithm. It is discussed in "The Role of the Step Size and Epsilon" on page 197.

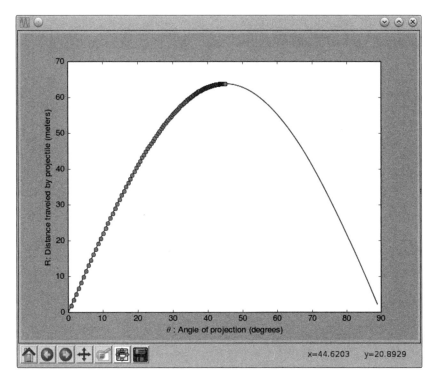

Figure 7-5: The gradient ascent method takes us iteratively toward the maximum point of the function.

The following grad_ascent() function implements the gradient ascent algorithm. The parameter x0 is the initial value of the variable at which to start the iteration, f1x is the derivative of the function whose maximum we want to find, and x is the Symbol object corresponding to the variable for the function.

```
'''
Use gradient ascent to find the angle at which the projectile
has maximum range for a fixed velocity, 25 m/s
'''

import math
from sympy import Derivative, Symbol, sin

def grad_ascent(x0, f1x, x):
❶    epsilon = 1e-6
❷    step_size = 1e-4
❸    x_old = x0
❹    x_new = x_old + step_size*f1x.subs({x:x_old}).evalf()
❺    while abs(x_old - x_new) > epsilon:
         x_old = x_new
         x_new = x_old + step_size*f1x.subs({x:x_old}).evalf()

    return x_new
```

```
❻ def find_max_theta(R, theta):
       # Calculate the first derivative
       R1theta = Derivative(R, theta).doit()
       theta0 = 1e-3
       theta_max = grad_ascent(theta0, R1theta, theta)
❼      return theta_max

   if __name__ == '__main__':

       g = 9.8
       # Assume initial velocity
       u = 25
       # Expression for range
       theta = Symbol('theta')
❽      R = u**2*sin(2*theta)/g

❾      theta_max = find_max_theta(R, theta)
       print('Theta: {0}'.format(math.degrees(theta_max)))
       print('Maximum Range: {0}'.format(R.subs({theta:theta_max})))
```

We set the epsilon value to 1e-6 and the step size to 1e-4 at ❶ and ❷, respectively. The epsilon value must always be a very small positive value close to 0, and the step size should be chosen such that the variable is incremented in small amounts at every iteration of the algorithm. The choice of the value of epsilon and step size is discussed in a bit more detail in "The Role of the Step Size and Epsilon" on page 197.

We set x_old to x0 at ❸ and calculate x_new for the first time at ❹. We use the subs() method to substitute the value of x_old in place of the variable and then use evalf() to calculate the numerical value. If the absolute difference abs(x_old - x_new) is greater than epsilon, the while loop at ❺ keeps executing, and we keep updating the value of x_old and x_new as per steps 1 and 2 of the gradient ascent algorithm. Once we're out of the loop—that is, abs(x_old - x_new) > epsilon—we return x_new, the variable value corresponding to the maximum function value.

We begin to define the find_max_theta() function at ❻. In this function, we calculate the first-order derivative of R; create a label, theta0, and set it to 1e-3; and call the grad_ascent() function with these two values as arguments, as well as a third argument, the symbol object theta. Once we get the value of θ corresponding to the maximum function value (theta_max), we return it at ❼.

Finally, we create the expression representing the horizontal range at ❽, having set the initial velocity, u = 25, and the theta Symbol object corresponding to the angle θ. Then we call the find_max_theta() function with R and theta at ❾.

When you run this program, you should see the following output:

```
Theta: 44.99999978475661
Maximum Range: 63.7755102040816
```

The value of θ is printed in degrees and turns out to be close to 45 degrees, as expected. If you change the initial velocity to other values, you'll see that the angle of projection at which the maximum range is reached is always close to 45 degrees.

A Generic Program for Gradient Ascent

We can modify the preceding program slightly to make a generic program for gradient ascent:

```
'''
Use gradient ascent to find the maximum value of a
single-variable function
'''

from sympy import Derivative, Symbol, sympify

def grad_ascent(x0, f1x, x):
    epsilon =  1e-6
    step_size = 1e-4
    x_old = x0
    x_new = x_old + step_size*f1x.subs({x:x_old}).evalf()
    while abs(x_old - x_new) > epsilon:
        x_old = x_new
        x_new = x_old + step_size*f1x.subs({x:x_old}).evalf()

    return x_new

if __name__ == '__main__':

    f = input('Enter a function in one variable: ')
    var = input('Enter the variable to differentiate with respect to: ')
    var0 = float(input('Enter the initial value of the variable: '))
    try:
        f = sympify(f)
    except SympifyError:
        print('Invalid function entered')
    else:
❶        var = Symbol(var)
❷        d = Derivative(f, var).doit()
❸        var_max = grad_ascent(var0, d, var)
        print('{0}: {1}'.format(var.name, var_max))
        print('Maximum value: {0}'.format(f.subs({var:var_max})))
```

The function grad_ascent() remains the same here. Now, however, the program asks the user to input the function, the variable in the function, and the initial value of the variable, where gradient ascent will begin. Once we're sure that SymPy can recognize the user's input, we create a Symbol object corresponding to the variable at ❶, find the first derivative with respect to it at ❷, and call the grad_ascent() function with these three arguments. The maximum value is returned at ❸.

Here's a sample run:

```
Enter a function in one variable: 25*25*sin(2*theta)/9.8
Enter the variable to differentiate with respect to: theta
Enter the initial value of the variable: 0.001
theta: 0.785360029379083
Maximum value: 63.7755100185965
```

The function input here is the same as in our first implementation of gradient ascent, and the value of θ is printed in radians.

Here's another run of the program, which will find the maximum value for cos$y$:

```
Enter a function in one variable: cos(y)
Enter the variable to differentiate with respect to: y
Enter the initial value of the variable: 0.01
y: 0.00999900001666658
Maximum value: 0.999950010415832
```

The program also works correctly for a function such as cos(y) + k, where k is a constant:

```
Enter a function in one variable: cos(y) + k
Enter the variable to differentiate with respect to: y
Enter the initial value of the variable: 0.01
y: 0.00999900001666658
Maximum value: k + 0.999950010415832
```

However, a function such as cos(ky) won't work because its first-order derivative, kcos(ky), still contains k, and SymPy doesn't know anything about its value. Therefore, SymPy can't perform a key step in the gradient ascent algorithm—namely, the comparison abs(x_old - x_new) > epsilon.

A Word of Warning About the Initial Value

The initial value of the variable from which we start the iteration of the gradient ascent method plays a very important role in the algorithm. Consider the function $x^5 - 30x^3 + 50x$, which we used as an example in Figure 7-3. Let's find the maximum using our generic gradient ascent program:

```
Enter a function in one variable: x**5 - 30*x**3 + 50*x
Enter the variable to differentiate with respect to: x
Enter the initial value of the variable: -2
x: -4.17445116397103
Maximum value: 705.959460322318
```

The gradient ascent algorithm stops when it finds the *closest peak*, which is not always the global maximum. In this example, when you start from the initial value of –2, it stops at the peak that also corresponds to the global maximum (approximately 706) in the considered domain. To verify this further, let's try a different initial value:

```
Enter a function in one variable: x**5 - 30*x**3 + 50*x
Enter the variable to differentiate with respect to: x
Enter the initial value of the variable: 0.5
x: 0.757452532565767
Maximum value: 25.0846622605419
```

In this case, the closest peak at which the gradient ascent algorithm stops is not the true global maximum of the function. Figure 7-6 depicts the result of the gradient ascent algorithm for both of these scenarios.

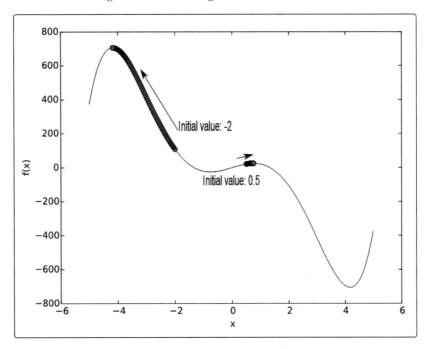

Figure 7-6: Results of the gradient ascent algorithm with different initial values. Gradient ascent always takes us to the closest peak.

Thus, when using this method, the initial value must be chosen carefully. Some variations of the algorithm try to address this limitation.

The Role of the Step Size and Epsilon

In the gradient ascent algorithm, the next value for the variable is calculated using the equation

$$\theta_{new} = \theta_{old} + \lambda \frac{dR}{d\theta},$$

where λ is the *step size*. The step size determines the distance of the next step. It should be small to avoid going *over* a peak. That is, if the current value of x is close to the value that corresponds to the maximum value of the function, the next step shouldn't be beyond the peak. The algorithm will then be unsuccessful. On the other hand, very small values will take longer to calculate. We've used a fixed step size of 10^{-3}, but this may *not* be the most appropriate value for all functions.

The value of epsilon (ε) that determines when we decide to stop the iteration of the algorithm should be a value that's sufficiently small that we're convinced the value of x is not changing. We expect the first derivative, $f'(x)$, to be 0 at the maximum point, and ideally the absolute difference $|\theta_{new} - \theta_{old}|$ is 0 (see step 2 of the gradient ascent algorithm on page 192). Due to numerical inaccuracies, however, we may not exactly get a difference of 0; hence, the value of epsilon is chosen to be a value close to 0, which, for all practical purposes, would tell us that the value of x isn't changing anymore. I have used 10^{-6} as the epsilon for all the functions. This value, although sufficiently small and suitable for the functions that have a solution for $f'(x) = 0$, such as sin(x), may not be the right value for other functions. Thus, it's a good idea to verify the maximum value at the end to ensure its correctness and, if needed, to adjust the value for epsilon accordingly.

Step 2 of the gradient ascent algorithm also implies that for the algorithm to terminate, the equation $f'(x) = 0$ must have a solution, which isn't the case for a function such as e^x or log(x). If you provide one of these functions as input to the preceding program, therefore, the program won't give you a solution, and it will continue running. We can make the gradient ascent program more useful for such cases by incorporating a check for whether $f'(x) = 0$ has a solution. Here's the modified program:

```
'''
Use gradient ascent to find the maximum value of a
single-variable function. This also checks for the existence
of a solution for the equation f'(x)=0.
'''

from sympy import Derivative, Symbol, sympify, solve

def grad_ascent(x0, f1x, x):
    # Check if f1x=0 has a solution
❶   if not solve(f1x):
        print('Cannot continue, solution for {0}=0 does not exist'.format(f1x))
        return
    epsilon = 1e-6
    step_size = 1e-4
    x_old = x0
    x_new = x_old + step_size*f1x.subs({x:x_old}).evalf()
    while abs(x_old - x_new) > epsilon:
        x_old = x_new
        x_new = x_old + step_size*f1x.subs({x:x_old}).evalf()

    return x_new
```

```
if __name__ == '__main__':

    f = input('Enter a function in one variable: ')
    var = input('Enter the variable to differentiate with respect to: ')
    var0 = float(input('Enter the initial value of the variable: '))
    try:
        f = sympify(f)
    except SympifyError:
        print('Invalid function entered')
    else:
        var = Symbol(var)
        d = Derivative(f, var).doit()
        var_max = grad_ascent(var0, d, var)
        if var_max:
            print('{0}: {1}'.format(var.name, var_max))
            print('Maximum value: {0}'.format(f.subs({var:var_max})))
```

❷

In this modification of the grad_ascent() function, we call SymPy's solve() function at ❶ to determine whether the equation $f'(x) = 0$, here f1x, has a solution. If not, we print a message and return. Another modification appears in the _main_ block at ❷. We check whether the grad_ascent() function successfully returned a result; if it did, then we proceed to print the maximum value of the function and the corresponding value of the variable.

These changes let the program handle functions such as $\log(x)$ and e^x:

```
Enter a function in one variable: log(x)
Enter the variable to differentiate with respect to: x
Enter the initial value of the variable: 0.1
Cannot continue, solution for 1/x=0 does not exist
```

You will see the same for e^x.

GRADIENT DESCENT ALGORITHM

The reverse algorithm of the gradient ascent algorithm is the gradient *descent* algorithm, which is a method to find the minimum value of a function. It is similar to the gradient ascent algorithm, but instead of "climbing up" along the function, we "climb down." Challenge #2 on page 205 discusses the difference between these two algorithms and gives you an opportunity to implement the reverse one.

Finding the Integrals of Functions

The *indefinite integral*, or the *antiderivative*, of a function $f(x)$ is another function $F(x)$, such that $F'(x) = f(x)$. That is, the integral of a function is another function whose derivative is the original function. Mathematically, it's written as $F(x) = \int f(x)\,dx$. The *definite integral*, on the other hand, is the integral

$$\int_a^b f(x)\,dx,$$

which is really $F(b) - F(a)$, where $F(b)$ and $F(a)$ are the values of the antiderivative of the function at $x = b$ and at $x = a$, respectively. We can find both the integrals by creating an Integral object.

Here's how we can find the integral $\int kx\,dx$, where k is a constant term:

```
>>> from sympy import Integral, Symbol
>>> x = Symbol('x')
>>> k = Symbol('k')
>>> Integral(k*x, x)
Integral(k*x, x)
```

We import the Integral and Symbol classes and create two Symbol objects corresponding to k and x. Then, we create an Integral object with the function kx, specifying the variable to integrate with respect to x. Similar to Limit and Derivative classes, we can now evaluate the integral using the doit() method:

```
>>> Integral(k*x, x).doit()
k*x**2/2
```

The integral turns out to be $kx^2/2$. If you calculate the derivative of $kx^2/2$, you'll get back the original function, kx.

To find the *definite* integral, we simply specify the variable, the lower limit, and the upper limit as a tuple when we create the Integral object:

```
>>> Integral(k*x, (x, 0, 2)).doit()
2*k
```

The result returned is the definite integral

$$\int_0^2 kx\,dx.$$

It can be useful to visualize definite integrals by discussing them in a geometric context. Consider Figure 7-7, which shows the graph of the function $f(x) = x$ between $x = 0$ and $x = 5$.

Now consider the region under the graph *ABDE*, which is bounded by the x-axis, between the points $x = 2$ and $x = 4$—points A and B, respectively. The area of the region can be found by adding the area of the square *ABCE* and the right-angled triangle *ECD*, which is $2 \times 2 + (1/2) \times 2 \times 2 = 6$.

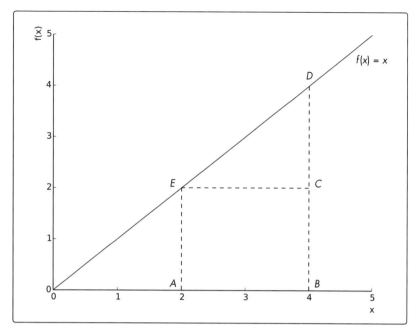

Figure 7-7: The definite integral of a function between two points is the area enclosed by the graph of the function bounded by the x-axis.

Let's now calculate the integral $\int_{2}^{4} x\,dx$:

```
>>> from sympy import Integral, Symbol
>>> x = Symbol('x')
>>> Integral(x, (x, 2, 4)).doit()
6
```

The value of the integral turns out to be the same as the area of the region *ABDE*. This isn't a coincidence; you'll find this is true for any function of *x* for which the integral can be determined.

Understanding that the definite integral is the area enclosed by the function between specified points on the *x*-axis is key for understanding probability calculations in random events that involve continuous random variables.

Probability Density Functions

Let's consider a fictional class of students and their grades on a math quiz. Each student can earn a grade between 0 and 20, including fractional grades. If we treat the grade as a random event, the grade itself is a *continuous random variable* because it can have *any* value between 0 and 20. If we want to calculate the probability of a student getting a grade between 11 and 12, we can't apply the strategy we learned in Chapter 5. To see why, let's consider the formula, assuming uniform probability,

$$P\left(11 < x < 12\right) = \frac{n\left(E\right)}{n\left(S\right)},$$

where E is the set of all grades possible between 11 and 12 and S is the set of all possible grades—that is, all real numbers between 1 and 20. By our definition of the preceding problem, $n(E)$ is infinite because it's impossible to count all possible real numbers between 11 and 12; the same is true for $n(S)$. Thus, we need a different approach to calculate the probability.

A *probability density function*, $P(x)$, expresses the probability of the value of a random variable being *close* to x, an arbitrary value.[1] It can also tell us the probability of x falling within an interval. That is, if we knew the probability density function representing the probability of grades in our fictional class, calculating $P(11 < x < 12)$ would give us the probability that we're looking for. But how do we calculate this? It turns out that this probability is the area enclosed by the graph of the probability density function and the x-axis between the points $x = 11$ and $x = 12$. Assuming an arbitrary probability density function, Figure 7-8 demonstrates this.

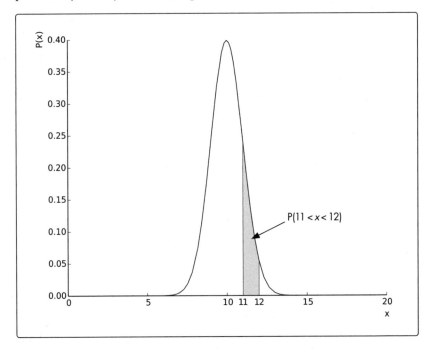

Figure 7-8: A probability density function for grades on a math quiz

1. For more information, see "The idea of a probability density function" by Duane Q. Nykamp from Math Insight (*http://mathinsight.org/probability_density_function_idea*).

We already know that this area is equal to the value of the integral,

$$\int_{11}^{12} p(x)\,dx;$$

thus, we have an easy way to find the probability of the grade lying between 11 and 12. With the math out of the way, we can now find out what the probability is. The probability density function we assumed earlier is the function

$$\frac{1}{\sqrt{2\pi}}e^{-\frac{(x-10)^2}{2}},$$

where x is the grade obtained. This function has been chosen so that the probability of the grade being close to 10 (either greater or less than) is high but then decreases sharply.

Now, let's calculate the integral

$$\int_{11}^{12} p(x)\,dx,$$

with $p(x)$ being the preceding function:

```
>>> from sympy import Symbol, exp, sqrt, pi, Integral
>>> x = Symbol('x')
>>> p = exp(-(x - 10)**2/2)/sqrt(2*pi)
>>> Integral(p, (x, 11, 12)).doit().evalf()
0.135905121983278
```

We create the Integral object for the function, with p representing the probability density function that specifies that we want to calculate the definite integral between 11 and 12 on the x-axis. We evaluate the function using doit() and find the numerical value using evalf(). Thus, the probability that a grade lies between 11 and 12 is close to 0.14.

THE PROBABILITY DENSITY FUNCTION: A CAVEAT

Strictly speaking, this density function assigns a nonzero probability to grades less than 0 or greater than 20. However, as you can check using the ideas from this section, the probability of such an event is so small that it is negligible for our purposes.

A probability density function has two special properties: (1) the function value for any x is always greater than 0, as probability can't be less than 0, and (2) the value of the definite integral

$$\int_{-\infty}^{\infty} f(x)\,dx$$

is equal to 1. The second property merits some discussion. Because $p(x)$ is a probability density function, the area enclosed by it, which is also the integral

$$\int_{a}^{b} p(x)\,dx,$$

between any two points, $x = a$ and $x = b$, gives us the probability of x lying between $x = a$ and $x = b$. This also means that no matter what the values of a and b are, the value of the integral must not exceed 1 because the probability can't be greater than 1 by definition. Hence, even if a and b are very large values such that they tend to $-\infty$ and ∞, respectively, the value of the integral will still be 1, as we can verify ourselves:

```
>>> from sympy import Symbol, exp, sqrt, pi, Integral, S
>>> x = Symbol('x')
>>> p = exp(-(x - 10)**2/2)/sqrt(2*pi)
>>> Integral(p, (x, S.NegativeInfinity, S.Infinity)).doit().evalf()
1.00000000000000
```

`S.NegativeInfinity` and `S.Infinity` denote the negative and positive infinity that we then specify as the lower and upper limits, respectively, while creating the `Integral` object.

When we're dealing with continuous random variables, a tricky situation can arise. In discrete probability, the probability of an event such as a fair six-sided die rolling a 7 is 0. We call an event for which the probability is 0 an *impossible* event. In the case of continuous random variables, the probability of the variable assuming any exact value is 0, even though it may be a *possible* event. For example, the grade of a student being exactly 11.5 is possible, but due to the nature of continuous random variables, the probability is 0. To see why, consider that the probability will be the value of the integral

$$\int_{11.5}^{11.5} p(x)\,dx.$$

Because this integral has the same lower and upper limits, its value is 0. This is rather unintuitive and paradoxical, so let's try to understand it.

Consider the range of grades we addressed earlier—0 to 20. The grade a student can obtain can be any number in this interval, which means there is an infinite number of numbers. If each number were to have an equal probability of being selected, what would that probability be? According to the formula for discrete probability, this should be $1/\infty$, which means a very small number. In fact, this number is so small that for all practical purposes, it's considered 0. Hence, the probability of the grade being 11.5 is 0.

What You Learned

In this chapter, you learned how to find the limits, derivatives, and integrals of functions. You learned about the gradient ascent method for finding the maximum value of a function and saw how you can apply integration principles to calculate the probability of continuous random variables. Next, you have a few tasks to attempt.

Programming Challenges

The following challenges build on what you've learned in this chapter. You can find sample solutions at *http://www.nostarch.com/doingmathwithpython/*.

#1: Verify the Continuity of a Function at a Point

A necessary, but not sufficient, condition for a function to be differentiable at a point is that it must be continuous at that point. That is, the function must be defined at that point and its left-hand limit and right-hand limit must exist and be equal to the value of the function at that point. If $f(x)$ is the function and $x = a$ is the point we are interested in evaluating, this is mathematically stated as

$$\lim_{x \to a^+} f(x) = \lim_{x \to a^-} f(x) = f(a).$$

Your challenge here is to write a program that will (1) accept a single-variable function and a value of that variable as inputs and (2) check whether the input function is continuous at the point where the variable assumes the value input.

Here is a sample working of the completed solution:

```
Enter a function in one variable: 1/x
Enter the variable: x
Enter the point to check the continuity at: 1
1/x is continuous at 1.0
```

The function $1/x$ is discontinuous at 0, so let's check that:

```
Enter a function in one variable: 1/x
Enter the variable: x
Enter the point to check the continuity at: 0
1/x is not continuous at 0.0
```

#2: Implement the Gradient Descent

The gradient descent method is used to find the minimum value of a function. Similar to the gradient ascent method, the gradient descent method is

an iterative method: we start with an initial value of the variable and gradually get closer to the variable value that corresponds to the minimum value
of the function. The step that gets us closer is the equation

$$x_{new} = x_{old} - \lambda \frac{df}{dx},$$

where λ is the step size and

$$\frac{df}{dx}$$

is the result of differentiating the function. Thus, the only difference
from the gradient ascent method is how we obtain the value of x_new from
x_old.

   Your challenge is to implement a generic program using the gradient
descent algorithm to find the minimum value of a single-variable function
specified as input by the user. The program should also create a graph
of the function and show all the intermediate values it found before finding the minimum. (You may want to refer to Figure 7-5 on page 193.)

#3: Area Between Two Curves

We learned that the integral

$$\int_a^b f(x)dx$$

expresses the area enclosed by the function $f(x)$, with the x-axis between
$x = a$ and $x = b$. The area between two curves is thus expressed as the
integral

$$\int_a^b \big(f(x) - g(x)\big)dx,$$

where a and b are the points of intersection of the two curves with $a < b$.
The function $f(x)$ is referred to as the *upper function* and $g(x)$ as the *lower
function*. Figure 7-9 illustrates this, assuming $f(x) = x$ and $g(x) = x^2$, with
$a = 0$ and $b = 1$.

   Your challenge here is to write a program that will allow the user to
input any two single-variable functions of x and print the enclosed area
between the two. The program should make it clear that the first function
entered should be the upper function, and it should also ask for the values
of x between which to find the area.

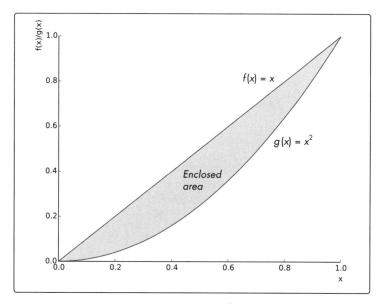

Figure 7-9: The functions f(x) = x and g(x) = x² enclose an area between x = 0 and x = 1.0.

#4: Finding the Length of a Curve

Let's say you just completed cycling along a road that looks roughly like Figure 7-10. Because you didn't have an odometer, you want to know whether there's a mathematical way to determine the distance you cycled. First, we'll need to find an equation—even an approximation will do— that describes this path.

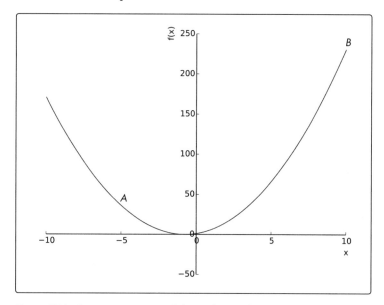

Figure 7-10: An approximation of the cycling path

Notice how it looks very similar to the quadratic functions we've discussed in the earlier chapters? In fact, for this challenge, let's assume that the equation is $y = f(x) = 2x^2 + 3x + 1$ and that you cycled from point A $(-5, 36)$ to point B $(10, 231)$. To find the length of this arc—that is, the distance you cycled—we'll need to calculate the integral

$$\int_a^b \sqrt{1 + \left(\frac{dy}{dx}\right)^2}\, dx,$$

where y describes the preceding function. Your challenge here is to write a program that will calculate the length of the arc, AB.

You may also want to generalize your solution so that it allows you to find the length of the arc between any two points for any arbitrary function, $f(x)$.

AFTERWORD

You've reached the end of the book—
nice work! Now that you've learned how
to handle numbers, generate graphs, apply
mathematical operations, manipulate sets and
algebraic expressions, create animated visualizations,
and solve calculus problems—whew!—what should
you do next? Here are a few things to try.[1]

Things to Explore Next

It's my hope that this book inspires you to go solve your own mathematical
questions. But it's often difficult to think up such challenges on your own.

---

1. For a clickable set of the links in this afterword, visit *http://nostarch.com/doingmathwithpython/*.

Project Euler

One definitive place to look for math problems that require you to implement programming solutions is Project Euler (*https://projecteuler.net/*), which offers more than 500 math problems of varying difficulty. Once you create a free account, you can submit your solutions to check whether they are correct.

Python Documentation

You may also wish to start exploring Python's documentation of various features.

- The Math Module: *https://docs.python.org/3/library/math.html*
- Other Numeric and Mathematical Modules: *https://docs.python.org/3/library/numeric.html*
- The Statistics Module: *https://docs.python.org/3/library/statistics.html*

We didn't discuss how floating point numbers are stored in a computer's memory or the problems and errors that may arise as a result. You may want to look at the decimal module's documentation and the discussion on "Floating Point Arithmetic" in the Python tutorial to learn about this topic:

- The Decimal Module: *https://docs.python.org/3/library/decimal.html*
- Floating Point Arithmetic: *https://docs.python.org/3.4/tutorial/floatingpoint.html*

Books

If you're interested in exploring more math and programming topics, check out the following books:

- *Invent Your Own Computer Games with Python* and *Making Games with Python and Pygame* by Al Sweigart (both freely available at *https://inventwithpython.com/*) don't specifically address solving math problems but apply math for the purpose of writing computer games using Python.
- *Think Stats: Probability and Statistics for Programmers* by Allen B. Downey is a freely available book (*http://greenteapress.com/thinkstats/*). As the title suggests, it delves deeply into statistics and probability topics beyond the ones discussed in this book.
- *Teach Your Kids to Code* by Bryson Payne (No Starch Press, 2015) is meant for beginners and covers various Python topics. You'll learn turtle graphics, various interesting ways of using the random Python module, and how to create games and animations using Pygame.

- *Computational Physics with Python* by Mark Newman (2013) focuses on a number of advanced math topics geared toward solving problems in physics. However, there are a number of chapters that are relevant to anyone interested in learning more about writing programs for solving numerical and mathematical problems.

Getting Help

If you are stuck on a specific issue discussed in this book, please contact me via email at *doingmathwithpython@gmail.com*. If you want to learn more about any of the functions or classes we have used in our programs, the first place to look would be the official documentation of the relevant projects:

- Python 3 standard library: *https://docs.python.org/3/library/index.html*
- SymPy: *http://docs.sympy.org/*
- matplotlib: *http://matplotlib.org/contents.html*

If you are stuck with a problem and want help, you can also email the project-specific mailing lists. You can find links to these on the book's website.

Conclusion

And finally, we've reached the end of the book. I hope you've learned a lot as you followed along. Go out there and solve some more problems using Python!

A

SOFTWARE INSTALLATION

 The programs and solutions in this book have been tested to run on Python 3.4, matplotlib 1.4.2, matplotlib-venn 0.11, and SymPy 0.7.6. These versions are only the minimum requirements, and the programs should also work with later versions of the software. Changes and updates will be noted on the book's website, *http://www.nostarch.com/doingmathwithpython/*.

While there are many ways to get your hands on Python and the libraries you need, one of the easiest is to use the Anaconda Python 3 software distribution, which is available freely for Microsoft Windows, Linux, and Mac OS X. At the time of this writing, the latest release of Anaconda is 2.1.0 with Python 3.4. Anaconda (*https://store.continuum.io/cshop/anaconda/*) is a quick and easy way to install Python 3 and many of the mathematical and data analysis packages, all in one easy installer. If you want to add new mathematical Python libraries, Anaconda also lets you add them quickly using the conda and pip commands. Anaconda has a number of other features that make it useful for Python development. It comes with the conda package manager built in, which allows the easy installation of third-party packages, as we'll soon see. It supports creating isolated Python environments, which means you can have multiple Python installations—for example, Python 2, Python 3.3, and Python 3.4—using the same Anaconda installation. You can learn more from the Anaconda website and the conda documentation (*http://conda.pydata.org/docs/intro.html*).

The next sections will briefly describe the installation of Anaconda on Microsoft Windows, Linux, and Mac OS X, so skip ahead to the section that applies to you. You'll need an Internet connection to follow along, but that's pretty much it.

If you run into any problems, there's also troubleshooting information available at *http://continuum.io/*.

Microsoft Windows

Download the Anaconda GUI installer for Python 3 from *http://continuum.io/downloads*. Double-click the installer and then follow these steps:

1. Click **Next** and accept the License Agreement:

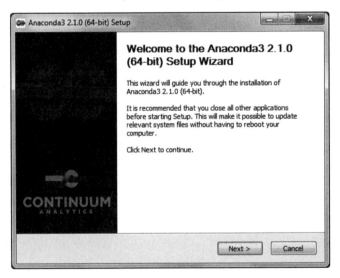

2. You can choose to install the distribution either for your username only or for all users using this computer.

3. Choose the folder where you want Anaconda to install the programs. The defaults should work fine.

4. Make sure to check the two boxes in the **Advanced Options** dialog so that you can invoke the Python shell and other programs, such as conda,

pip, and idle, from anywhere on the command prompt. In addition, any other Python programs looking for a Python 3.4 installation will be pointed to the one installed by Anaconda:

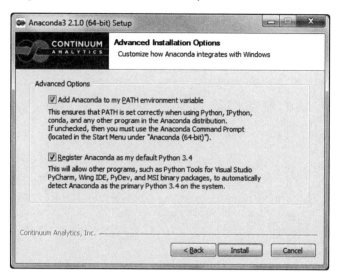

5. Click **Install** to start the installation. When the installation has finished, click **Next** and then click **Finish** to complete the installation. You should be able to find Python in your Start Menu.

6. Open a Windows command prompt and carry out the following steps.

Updating SymPy

The installation may come with SymPy already installed, but we want to make sure that we have at least 0.7.6, so we'll install it using this command:

```
$ conda install sympy=0.7.6
```

This will install or upgrade to SymPy 0.7.6.

Installing matplotlib-venn

To install matplotlib-venn, use this command:

```
$ pip install matplotlib-venn
```

Your computer is now set up to run all the programs.

Starting the Python Shell

Open a Windows command prompt and enter idle to start the IDLE shell or python to start the Python 3 default shell.

Linux

The Linux installer is distributed as a shell script installer, so you'll want to download the Anaconda Python installer from *http://continuum.io/downloads*. Then start the installer by executing the following:

```
$ bash Anaconda3-2.1.0-Linux-x86_64.sh

Welcome to Anaconda3 2.1.0 (by Continuum Analytics, Inc.)

In order to continue the installation process, please review the license
agreement.
Please, press ENTER to continue
>>>
```

The "Anaconda END USER LICENSE AGREEMENT" will be displayed. Once you've read through it, enter **yes** to continue the installation:

```
Do you approve the license terms? [yes|no]
[no] >>> yes

Anaconda3 will now be installed into this location:
/home/testuser/anaconda3

  - Press ENTER to confirm the location
  - Press CTRL-C to abort the installation
  - Or specify a different location below
```

Press ENTER at the prompt, and the installation will start:

```
[/home/testuser/anaconda3] >>>
PREFIX=/home/testuser/anaconda3
installing: python-3.4.1-4 ...
installing: conda-3.7.0-py34_0
..

creating default environment...
installation finished.
Do you wish the installer to prepend the Anaconda3 install location
to PATH in your /home/testuser/.bashrc ? [yes|no]
```

When asked to confirm the install location, enter **yes** so that the Python 3.4 interpreter installed by Anaconda is always invoked when you invoke the Python program from your terminal:

```
[no] >>> yes

Prepending PATH=/home/testuser/anaconda3/bin to PATH in /home/testuser/.bashrc
A backup will be made to: /home/testuser/.bashrc-anaconda3.bak
```

For this change to become active, you have to open a new terminal.

Thank you for installing Anaconda3!

Open a new terminal for the next steps.

Updating SymPy

First, make sure that SymPy 0.7.6 is installed:

```
$ conda install sympy=0.7.6
```

Installing matplotlib-venn

Use the following command to install matplotlib-venn:

```
$ pip install matplotlib-venn
```

Starting the Python Shell

You're all set. Open a new terminal and enter idle3 to start the IDLE editor or python to start the Python 3.4 shell. You should now be able to run all the programs and try out new ones.

Mac OS X

Download the graphical installer from *http://continuum.io/downloads*. Then double-click the *.pkg* file and follow the instructions:

1. Click **Continue** on each informational window:

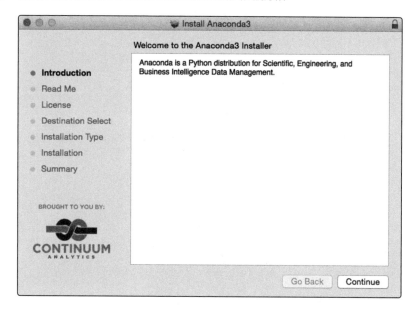

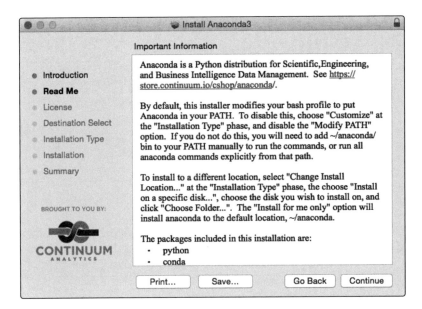

2. Click **Agree** to accept the "Anaconda END USER LICENSE AGREEMENT":

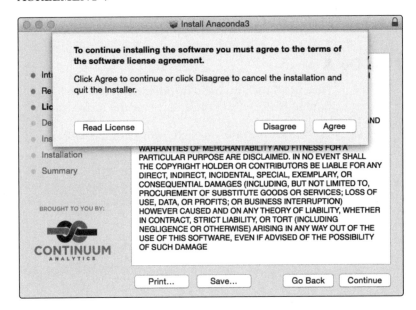

3. In the following dialog, choose the "Install for me only" option. The error message you see is a bug in the installer software. Just click it, and it will disappear. Click **Continue** to proceed.

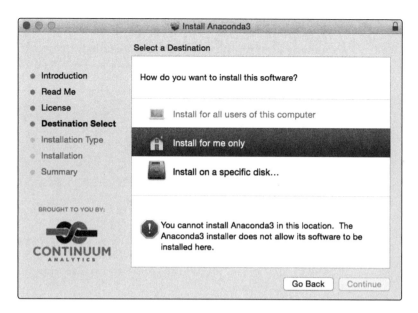

4. Select **Install**:

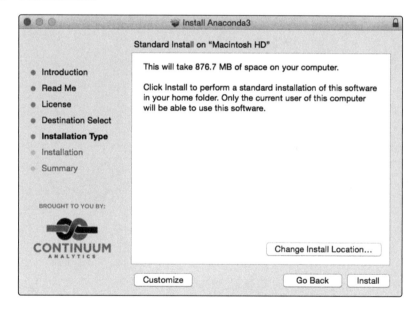

5. Once the installation is finished, open the Terminal app and follow the next steps to update SymPy and install matplotlib-venn.

Updating SymPy

First, make sure that SymPy 0.7.6 is installed:

```
$ conda install sympy=0.7.6
```

Installing matplotlib-venn

Use the following command to install matplotlib-venn:

```
$ pip install matplotlib-venn
```

Starting the Python Shell

You're all set. Close the Terminal window, open a new one, and enter idle3 to start the IDLE editor or python to start the Python 3.4 shell. You should now be able to run all the programs and try out new ones.

B

OVERVIEW OF PYTHON TOPICS

The aim of this appendix is twofold: to provide a quick refresher on some Python topics that weren't thoroughly introduced in the chapters and to introduce topics that will help you write better Python programs.

if __name__ == '__main__'

Throughout the book, we've used the following block of code, where func() is a function we've defined in the program:

```
if __name__ == '__main__':
    # Do something
    func()
```

This block of code ensures that the statements within the block are executed only when the program is run on its own.

When a program runs, the special variable __name__ is set to __main__ automatically, so the if condition evaluates to True and the function func() is called. However, __name__ is set differently when you import the program into another program (see "Reusing Code" on page 235).

Here's a quick demonstration. Consider the following program, which we'll call *factorial.py*:

```
# Find the factorial of a number
def fact(n):
    p = 1
    for i in range(1, n+1):
        p = p*i
    return p

❶ print(__name__)

if __name__ == '__main__':
    n = int(input('Enter an integer to find the factorial of: '))
    f = fact(n)
    print('Factorial of {0}: {1}'.format(n, f))
```

The program defines a function, fact(), that calculates the factorial of the integer passed to it. When you run it, it prints __main__, which corresponds to the print statement at ❶, because __name__ is automatically set to __main__. Then, it asks an integer to be entered, calculates the factorial, and prints it:

```
__main__
Enter an integer to find the factorial of: 5
Factorial of 5: 120
```

Now, say you need to calculate the factorial in another program. Instead of writing the function again, you decide to reuse this function by importing it:

```
from factorial import fact
if __name__ == '__main__':
    print('Factorial of 5: {0}'.format(fact(5)))
```

Note that both the programs must be in the same directory. When you run this program, you'll get the following output:

```
factorial
Factorial of 5: 120
```

When your program is imported by another program, the value of the variable __main__ is set to that program's filename, without the extension. In this case, the value of __name__ is factorial instead of __main__. Because the condition __name__ == '__main__' now evaluates to False, the program doesn't ask for the user's input anymore. Remove the condition to see for yourself what happens!

To summarize, it's good practice to use if __name__ == '__main__' in your programs so that the statements you want executed when your program is run as a standalone are also *not* executed when your program is imported into another program.

List Comprehensions

Let's say we have a list of integers and we want to create a new list containing the squares of the elements of the original list. Here's one way that we could do this that's already familiar to you:

```
>>> x = [1, 2, 3, 4]
>>> x_square = []
❶ >>> for n in x:
❷        x_square.append(n**2)
>>> x_square
[1, 4, 9, 16]
```

Here, we used a code pattern that we've used in various programs throughout the book. We create an empty list, x_square, and then successively append to it as we calculate the square. We can do this in a more efficient way using *list comprehensions*:

```
❸ >>> x_square = [n**2 for n in x]
>>> x_square
[1, 4, 9, 16]
```

The statement at ❸ is referred to as a *list comprehension* in Python. It consists of an expression—here, n**2—followed by a for loop, for n in x. Note that it basically allows us to combine the two statements at ❶ and ❷ into one to create a new list in one statement.

As another example, consider one of the programs we wrote in "Drawing the Trajectory" on page 51 to draw the trajectory of a body in projectile motion. In these programs, we have the following block of code to calculate the *x*- and *y*-coordinates of the body at each time instant:

```
# Find time intervals
intervals = frange(0, t_flight, 0.001)
# List of x and y coordinates
x = []
y = []
for t in intervals:
    x.append(u*math.cos(theta)*t)
    y.append(u*math.sin(theta)*t - 0.5*g*t*t)
```

Using list comprehension, you can rewrite the block of code as follows:

```
# Find time intervals
intervals = frange(0, t_flight, 0.001)
# List of x and y coordinates
```

```
x = [u*math.cos(theta)*t for t in intervals]
y = [u*math.sin(theta)*t - 0.5*g*t*t for t in intervals]
```

The code is more compact now, as you didn't have to create the empty lists, write a for loop, and append to the lists. List comprehension lets you do this in a single statement.

You can also add conditionals to a list comprehension in order to selectively choose which list items are evaluated in the expression. Consider, once again, the first example:

```
>>> x = [1, 2, 3, 4]
>>> x_square = [n**2 for n in x if n%2 == 0]
>>> x_square
[4, 16]
```

In this list comprehension, we use the if condition to explicitly tell Python to evaluate the expression n**2 only on the even list items of x.

Dictionary Data Structure

We first used a Python dictionary in Chapter 4 while implementing the subs() method in SymPy. Let's explore Python dictionaries in more detail. Consider a simple dictionary:

```
>>> d = {'key1': 5, 'key2': 20}
```

This code creates a dictionary with two keys—'key1' and 'key2'—with values 5 and 20, respectively. Only strings, numbers, and tuples can be keys in a Python dictionary. These data types are referred to as *immutable* data types—once created, they can't be changed—so a list can't be a key because we can add and remove elements from a list.

We already know that to retrieve the value corresponding to 'key1' in the dictionary, we need to specify it as d['key1']. This is one of the most common use cases of a dictionary. A related use case is checking whether the dictionary contains a certain key, 'x'. We can check that as follows:

```
>>> d = {'key1': 5, 'key2': 20}
>>> 'x' in d
False
```

Once we create a dictionary, we can add a new key-value pair to it, similar to how we can append elements to a list. Here's an example:

```
>>> d = {'key1': 5, 'key2': 20}
>>> if 'x' in d:
        print(d['x'])
else:
        d['x'] = 1
```

```
>>> d
{'key1': 5, 'x': 1, 'key2': 20}
```

This code snippet checks whether the key 'x' already exists in the dictionary, d. If it does, it prints the value corresponding to it; otherwise, it adds the key to the dictionary with 1 as the corresponding value. Similar to Python's behavior with sets, Python can't guarantee a particular order of the key-value pairs in a dictionary. The key-value pairs can be in any order, irrespective of the order of insertion.

Besides specifying the key as an index to the dictionary, we can also use the get() method to retrieve the value corresponding to the key:

```
>>> d.get('x')
1
```

If you specify a nonexistent key to the get() method, None is returned. On the other hand, if you do so while using the index style of retrieving, you'll get an error.

The get() method also lets you set a default value for nonexistent keys:

```
>>> d.get('y', 0)
0
```

There's no key 'y' in the dictionary d, so 0 is returned. If there is a key, however, the value is returned instead:

```
>>> d['y'] = 1
>>> d.get('y', 0)
1
```

The keys() and values() methods each return a list-like data structure of all the keys and values, respectively, in a dictionary:

```
>>> d.keys()
dict_keys(['key1', 'x', 'key2', 'y'])
>>> d.values()
dict_values([5, 1, 20, 1])
```

To iterate over the key and value pairs in a dictionary, use the items() method:

```
>>> d.items()
dict_items([('key1', 5), ('x', 1), ('key2', 20), ('y', 1)])
```

This method returns a *view* of tuples, and each tuple is a key-value pair. We can use the following code snippet to print them nicely:

```
>>> for k, v in d.items():
        print(k, v)
```

```
key1 5
x 1
key2 20
y 1
```

Views are more memory efficient than lists, and they don't let you add or remove items.

Multiple Return Values

In the programs we've written so far, most of the functions return a single value, but functions sometimes return multiple values. We saw an example of such a function in "Measuring the Dispersion" on page 71, where in the program to find the range, we returned three numbers from the find_range() function. Here's another example of the approach we took there:

```
import math
def components(u, theta):
    x = u*math.cos(theta)
    y = u*math.sin(theta)
    return x, y
```

The components() function accepts a velocity, u, and an angle, theta, in radians as parameters, and it calculates the x and y components and returns them. To return the calculated components, we simply list the corresponding Python labels in the return statement separated by a comma. This creates and returns a tuple consisting of the items x and y. In the calling code, we receive the multiple values:

```
if __name__ == '__main__':
    theta = math.radians(45)
    x, y = components(theta)
```

Because the components() function returns a tuple, we can retrieve the returned values using tuple indices:

```
c = components(theta)
x = c[0]
y = c[1]
```

This has advantages because we don't have to know all the different values being returned. For one, you don't have to write x,y,z = myfunc1() when the function returns three values or a,x,y,z = myfunc1() when the function returns four values, and so on.

In either of the preceding cases, the code calling the components() function must know which of the return values correspond to which component of the velocity, as there's no way to know that from the values themselves.

A user-friendly approach is to return a dictionary object instead, as we saw in the case of SymPy's solve() function when used with the dict=True keyword argument. Here's how we can rewrite the preceding components function to return a dictionary:

```python
import math

def components(theta):
    x = math.cos(theta)
    y = math.sin(theta)

    return {'x': x, 'y': y}
```

Here, we return a dictionary with the keys 'x' and 'y' referring to the x and y components and their corresponding numerical values. With this new function definition, we don't need to worry about the order of the returned values. We just use the key 'x' to retrieve the x component and the key 'y' to retrieve the y component:

```python
if __name__ == '__main__':
    theta = math.radians(45)
    c = components(theta)
    y = c['y']
    x = c['x']
    print(x, y)
```

This approach eliminates the need to use indices to refer to a specific returned value. The following code rewrites the program to find the range (see "Measuring the Dispersion" on page 71) so that the results are returned as a dictionary instead of a tuple:

```python
'''
Find the range using a dictionary to return values
'''
def find_range(numbers):
    lowest = min(numbers)
    highest = max(numbers)
    # Find the range
    r = highest-lowest
    return {'lowest':lowest, 'highest':highest, 'range':r}

if __name__ == '__main__':
    donations = [100, 60, 70, 900, 100, 200, 500, 500, 503, 600, 1000, 1200]
    result = find_range(donations)
❶  print('Lowest: {0} Highest: {1} Range: {2}'.
          format(result['lowest'], result['highest'], result['range']))
```

The find_range() function now returns a dictionary with the keys lowest, highest, and range and with the lowest number, highest number, and the range as their corresponding values. At ❶, we simply use the corresponding key to retrieve the corresponding value.

If we were just interested in the range of a group of numbers and we didn't care about the lowest and highest numbers, we'd just use result['range'] and not worry about what other values were returned.

Exception Handling

In Chapter 1, we learned that trying to convert a string such as '1.1' to an integer using the int() function results in a ValueError exception. But with a try...except block, we can print a user-friendly error message:

```
>>> try:
        int('1.1')
except ValueError:
        print('Failed to convert 1.1 to an integer')

Failed to convert 1.1 to an integer
```

When any statement in the try block raises an exception, the type of exception raised is matched with the one specified by the except statement. If there's a match, the program resumes in the except block. If the exception doesn't match, the program execution halts and displays the exception. Here's an example:

```
>>> try:
        print(1/0)
except ValueError:
        print('Division unsuccessful')

Traceback (most recent call last):
  File "<pyshell#66>", line 2, in <module>
    print(1/0)
ZeroDivisionError: division by zero
```

This code block attempts a division by 0, which results in a ZeroDivisionError exception. Although the division is carried out in a try...except block, the exception type is incorrectly specified, and the exception isn't handled correctly. The correct way to handle this exception is to specify ZeroDivisionError as the exception type.

Specifying Multiple Exception Types

You can also specify multiple exception types. Consider the function reciprocal(), which returns the reciprocal of the number passed to it:

```
def reciprocal(n):
    try:
        print(1/n)
    except (ZeroDivisionError, TypeError):
        print('You entered an invalid number')
```

We defined the function reciprocal(), which prints the reciprocal of the user's input. We know that if the function is called with 0, it'll cause a ZeroDivisionError exception. If you pass a string, however, it'll cause a TypeError exception. The function considers both these cases as invalid input and specifies both ZeroDivisionError and TypeError in the except statement as a tuple.

Let's try calling the function with a valid input—that is, a nonzero number:

```
>>> reciprocal(5)
0.2
```

Next, we call the function with 0 as the argument:

```
>>> reciprocal(0)
Enter an integer: 0
You entered an invalid number
```

The 0 argument raises the ZeroDivisionError exception, which is in the tuple of exception types specified to the except statement, so the code prints an error message.

Now, let's enter a string:

```
>>> reciprocal('1')
```

In this case, we entered an invalid number, which raises the TypeError exception. This exception is also in the tuple of specified exceptions, so the code prints an error message. If you want to give a more specific error message, we can just specify multiple except statements as follows:

```
def reciprocal(n):
    try:
        print(1/n)
    except TypeError:
        print('You must specify a number')
    except ZeroDivisionError:
        print('Division by 0 is invalid')

>>> reciprocal(0)
Division by 0 is invalid
>>> reciprocal('1')
You must specify a number
```

In addition to TypeError, ValueError, and ZeroDivisionError, there are a number of other built-in exception types. The Python documentation at *https://docs.python.org/3.4/library/exceptions.html#bltin-exceptions* lists the built-in exceptions for Python 3.4.

The else Block

The else block is used to specify which statements to execute when there's no exception. Consider an example from the program we wrote to draw the trajectory of a projectile (see "Drawing the Trajectory" on page 51):

```
if __name__ == '__main__':
    try:
        u = float(input('Enter the initial velocity (m/s): '))
        theta = float(input('Enter the angle of projection (degrees): '))
    except ValueError:
        print('You entered an invalid input')
❶   else:
        draw_trajectory(u, theta)
        plt.show()
```

If the input for u or theta couldn't be converted to a floating point number, it doesn't make sense for the program to call the draw_trajectory() and plt.show() functions. Instead, we specify these two statements in the else block at ❶. Using try...except...else will let you manage different types of errors during runtime and take appropriate action when there is an error or when there is none:

1. If there's an exception and there's an except statement corresponding to the exception type raised, the execution is transferred to the corresponding except block.

2. If there's no exception, the execution is transferred to the else block.

Reading Files in Python

Opening a file is the first step to reading data from it. Let's start with a quick example. Consider a file that consists of a collection of numbers with one number per line:

```
100
60
70
900
100
200
500
500
503
600
1000
1200
```

We want to write a function that reads the file and returns a list of those numbers:

```
def read_data(path):
    numbers = []
❶   f = open(path)
❷   for line in f:
        numbers.append(float(line))
    f.close()
    return numbers
```

First, we define the function read_data() and create an empty list to store all of the numbers. At ❶, we use the open() function to open the file whose location has been specified via the argument path. An example of the path would be */home/username/mydata.txt* on Linux, *C:\mydata.txt* on Microsoft Windows, or */Users/Username/mydata.txt* on OS X. The open() function returns a file object, which we use the label f to refer to. We can go over each line of the file using a for loop at ❷. Because each line is returned as a string, we convert it into a number and append it to the list numbers. The loop stops executing once all the lines have been read, and we close the file using the close() method. Finally, we return the numbers list.

This is similar to how we read the numbers from a file in Chapter 3, although we didn't have to close the file explicitly because we used a different approach there. Using the approach we took in Chapter 3, we would rewrite the preceding function as follows:

```
def read_data(path):
    numbers = []
❶   with open(path) as f:
        for line in f:
            numbers.append(float(line))
❷   return numbers
```

The key statement here is at ❶. It's similar to writing f = open(path) but only partially. Besides opening the file and assigning the file object returned by open() to f, it also sets up a new *context* with all the statements in that block—in this case, all the statements before the return statement. When all the statements in the body have been executed, the file is automatically closed. That is, when the execution reaches the statement at ❷, the file is closed without needing an explicit call to the close() method. This method also means that if there are any exceptions while working with the file, it'll still be closed before the program exits. This is the preferred approach to working with files.

Reading All the Lines at Once

Instead of reading the lines one by one to build a list, we can use the readlines() method to read all the lines into a list at once. This results in a more compact function:

```
def read_data(path):
    with open(path) as f:
❶       lines = f.readlines()
    numbers = [float(n) for n in lines]
    return numbers
```

We read all the lines of the file into a list using the readlines() method at ❶. Then, we convert each of the items in the list into a floating point number using the float() function and list comprehension. Finally, we return the list numbers.

Specifying the Filename as Input

The read_data() function takes the file path as an argument. If your program allows you to specify the filename as an input, this function should work for any file as long as the file contains data we expect to read. Here's an example:

```
if __name__=='__main__':
    data_file = input('Enter the path of the file: ')
    data = read_data(data_file)
    print(data)
```

Once you've added this code to the end of the read_data() function and run it, it'll ask you to input the path to the file. Then, it'll print the numbers it reads from the file:

```
Enter the path of the file: /home/amit/work/mydata.txt
[100.0, 60.0, 70.0, 900.0, 100.0, 200.0, 500.0, 500.0, 503.0, 600.0, 1000.0,
1200.0]
```

Handling Errors When Reading Files

There are a couple of things that can go wrong when reading files: (1) the file can't be read, or (2) the data in the file isn't in the expected format. Here's an example of what happens when a file can't be read:

```
Enter the path of the file: /home/amit/work/mydata2.txt
Traceback (most recent call last):
  File "read_file.py", line 11, in <module>
    data = read_data(data_file)
  File "read_file.py", line 4, in read_data
    with open(path) as f:
FileNotFoundError: [Errno 2] No such file or directory: '/home/amit/work/
mydata2.txt'
```

Because I entered a file path that doesn't exist, the `FileNotFoundError` exception is raised when we try to open the file. We can make the program display a user-friendly error message by modifying our read_data() function as follows:

```
def read_data(path):
    numbers = []
    try:
        with open(path) as f:
            for line in f:
                numbers.append(float(line))
    except FileNotFoundError:
        print('File not found')
    return numbers
```

Now, when you specify a nonexistent file path, you'll get an error message instead:

```
Enter the path of the file: /home/amit/work/mydata2.txt
File not found
```

The second source of errors can be that the data in the file isn't what your program expects to read. For example, consider a file that has the following:

```
10
20
3o
1/5
5.6
```

The third line in this file isn't convertible to a floating point number because it has the letter o in it instead of the number 0, and the fourth line consists of 1/5, a fraction in string form, which float() can't handle.

If you supply this data file to the earlier program, it'll produce the following error:

```
Enter the path of the file: bad_data.txt
Traceback (most recent call last):
  File "read_file.py", line 13, in <module>
    data = read_data(data_file)
  File "read_file.py", line 6, in read_data
    numbers.append(float(line))
ValueError: could not convert string to float: '3o\n'
```

The third line in the file is 3o, not the number 30, so when we attempt to convert it into a floating point number, the result is ValueError. There are two approaches you can take when such data is present in a file. The first is to report the error and exit the program. The modified read_data() function would appear as follows:

```python
def read_data(path):
    numbers = []
    try:
        with open(path) as f:
            for line in f:
❶              try:
❷                  n = float(line)
                except ValueError:
                    print('Bad data: {0}'.format(line))
❸                  break
❹              numbers.append(n)
    except FileNotFoundError:
        print('File not found')
    return numbers
```

We insert another try...except block in the function starting at ❶, and we convert the line into a floating point number at ❷. If the program raises the ValueError exception, we print an error message with the offending line and exit out of the for loop using break at ❸. The program then stops reading the file. The returned list, numbers, contains all the data that was successfully read before encountering the bad data. If there's no error, we append the floating point number to the numbers list at ❹.

Now when you supply the file *bad_data.txt* to the program, it'll read only the first two lines, display the error message, and exit:

```
Enter the path of the file: bad_data.txt
Bad data: 30

[10.0, 20.0]
```

Returning partial data may not be desirable, so we could just replace the break statement at ❸ with return and no data would be returned.

The second approach is to ignore the error and continue with the rest of the file. Here's a modified read_data() function that does this:

```python
def read_data(path):
    numbers = []
    try:
        with open(path) as f:
            for line in f:
                try:
                    n = float(line)
                except ValueError:
                    print('Bad data: {0}'.format(line))
❶                  continue
                numbers.append(n)
    except FileNotFoundError:
        print('File not found')
    return numbers
```

The only change here is that instead of breaking out of the for loop, we just continue with the next iteration using the continue statement at ❶. The output from the program is now as follows:

```
Bad data: 30

Bad data: 1/5

[10.0, 20.0, 5.6]
```

The specific application where you're reading the file will determine which of the above approaches you want to take to handle bad data.

Reusing Code

Throughout this book, we've used classes and functions that were either part of the Python standard library or available after installing third-party packages, such as matplotlib and SymPy. Now we'll look at a quick example of how we can import our own programs into other programs.

Consider the function find_corr_x_y() that we wrote in "Calculating the Correlation Between Two Data Sets" on page 75. We'll create a separate file, *correlation.py*, which has only the function definition:

```python
'''
Function to calculate the linear correlation coefficient
'''

def find_corr_x_y(x,y):
    # Size of each set
    n = len(x)

    # Find the sum of the products
    prod=[]
    for xi,yi in zip(x,y):
        prod.append(xi*yi)

    sum_prod_x_y = sum(prod)
    sum_x = sum(x)
    sum_y = sum(y)
    squared_sum_x = sum_x**2
    squared_sum_y = sum_y**2

    x_square=[]
    for xi in x:
        x_square.append(xi**2)
    x_square_sum = sum(x_square)

    y_square=[]
    for yi in y:
        y_square.append(yi**2)
    y_square_sum = sum(y_square)
```

```
numerator = n*sum_prod_x_y - sum_x*sum_y
denominator_term1 = n*x_square_sum - squared_sum_x
denominator_term2 = n*y_square_sum - squared_sum_y
denominator = (denominator_term1*denominator_term2)**0.5

correlation = numerator/denominator

return correlation
```

Without the *.py* file extension, a Python file is referred to as a module. This is usually reserved for files that define classes and functions that'll be used in other programs. The following program imports the find_corr_x_y() function from the correlation module we just defined:

```
from correlation import find_corr_x_y
if __name__ == '__main__':
    high_school_math = [83, 85, 84, 96, 94, 86, 87, 97, 97, 85]
    college_admission = [85, 87, 86, 97, 96, 88, 89, 98, 98, 87]
    corr = find_corr_x_y(high_school_math, college_admission)
    print('Correlation coefficient: {0}'.format(corr))
```

This program finds the correlation between the high school math grades and college admission scores of students we considered in Table 3-3 on page 80. We import the find_corr_x_y() function from the correlation module, create the lists representing the two sets of grades, and call the find_corr_x_y() function with the two lists as arguments. When you run the program, it'll print the correlation coefficient. Note that the two files must be in the same directory—this is strictly to keep things simple.

INDEX

in operator, 122
input() function, 8
installation, of software
    on Linux, 216–217
    on Mac OS X, 217–220
    on Windows, 214–215
Integral class, 200
integrals of functions, finding, 200
intersection, of sets, 127
interval argument, 154

K

keys, in a dictionary, 224, 227

L

labels, 4
Lady ferns, 164
law of large numbers, 144
legend() function, 40
len() function, 62
limit, finding, 181
Limit class, 182
Linux, software installation on,
    216–217
lists, 29–31
    appending to a list, 30
    choosing a random
        element, 161
    creating a set, 123
    empty lists, 30
    index, 29
    iterating over the elements, 31
    len() function, 62
    list comprehensions, 223–224
    lists of lists, 173–175
    max() function, 72
    min() function, 72
    sort() method, 64
    sum() function, 62
    tuples as members, 66
    zip() function, 77
local maxima and minima,
    188–191
log() function, 179

M

Mac OS X, software installation on,
    217–220
Mandelbrot set, 172–176
mathematical operations, 1–3
    exponential operator, 3
    floor division operator, 2
    modulo (%) operator, 3, 12
math module, 178
matplotlib, 32
    animation module, 154
    axes
        auto scaling, 152
        customizing, 42
    Axes object, 151
    axis() function, 43
    barh() function, 57
    Circle patch, 151
    colorbar() function, 175
    displaying images, 172
    documentation, 211
    Figure object, 150, 154
    FuncAnimation class, 154–158
    gca() function, 152
    gcf() function, 154
    imshow() function, 172
    labels, 41
    legend, adding a, 40
    legend() function, 40
    marker, 34
    multiple data sets, 38, 53
    patches, 150
    plot() function, 32, 36
    Polygon patch, 168
    pylab module, 32
    pyplot module, 44
    savefig() function, 45
    saving, 45–46
    scatter() function, 81
    scatter plots, 79, 81–83
    set_aspect() method, 153
    show() function, 32
    title, 41
    title() function, 41
    xlabel() function, 41
    ylabel() function, 41

maxima and minima, of functions, 188–191
max() function, 72
mean, finding, 62–63
median, finding, 63–65
min() function, 72
mode, finding, 65–69
modules, 5
modulo (%) operator, 3
multiplication tables, generating, 15–17, 23
multiplying expressions, 104–105

N

__name__, 221–223
negative index, of a list, 31
NegativeInfinity, 204
Newton's law of universal gravitation, 46–48
number line, 28
numbers
    abs() function, 7
    common number sets, 126
    complex numbers. *See* complex numbers
    conversion between types, 5
    float() function, 5
    floating point, 4–5
    Fraction class, 5, 6
    fractions module, 5
    integers, 4–5
    int() function, 5
    is_integer() method, 10
    random. *See* random numbers
    rational, irrational, and real, 126
    type() function, 4
    types of, 4–7
Nykamp, Duane Q., "The idea of a probability density function," 202

O

open() function, 231
order of operations (PEMDAS), 3

P

Packages (Python), 32
partial derivative of functions, finding, 187
Pearson correlation coefficient, 75
PEMDAS (order of operations), 3
pi (π), estimating value of, 147
plot() function, 32, 109
plotting
    expressions, 108–115
        input by the user, 111–113
        multiple, 113–115
    with formulas, 46–54
    projectile motion, 48–54
    using SymPy. *See* SymPy
polynomial expressions, 117
polynomial() method, 119
pretty printing, 97–100
probability, 131–140, 201–204
    continuous random variable, 201
    density functions, 201–204
    distribution, uniform, 131
    expectation, 143
    law of large numbers, 144
    nonuniform probability, 164
    random numbers. *See also* random numbers
        generating, 134–137
        nonuniform, 137–140
    random variable, 143
Project Euler, 210
projectile motion, 48, 191
    animation, 156
    trajectory drawing, 51, 56
pylab module, 32
pyplot module, 44–45
Python
    documentation, 210, 211
    IDLE, 1, 13–14
    installation
        Linux, 216–217
        Mac OS X, 217–220
        Windows, 214–215
    overview, 221–236

W

while loop, 24
    exiting early using break, 24
Windows, software installation on,
      214–215

Z

ZeroDivisionError, 11, 228–229
zip() function, 77

Doing Math with Python is set in New Baskerville, Futura, Dogma, and TheSansMono Condensed. The book was printed and bound by Lake Book Manufacturing in Melrose Park, Illinois. The paper is 60# Husky Opaque Offset, which is certified by the Sustainable Forestry Initiative (SFI).

The book uses a layflat binding, in which the pages are bound together with a cold-set, flexible glue, and the first and last pages of the resulting book block are attached to the cover. The cover is not actually glued to the book's spine, and when open, the book lies flat and the spine doesn't crack.

RESOURCES

Visit *https://www.nostarch.com/doingmathwithpython/* for resources, errata, and more information.

More no-nonsense books from **NO STARCH PRESS**

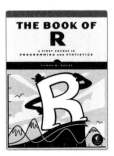

THE BOOK OF R
A First Course in Programming and Statistics
by TILMAN M. DAVIES
FALL 2015, 600 PP., $39.95
ISBN 978-1-59327-651-5

STATISTICS DONE WRONG
The Woefully Complete Guide
by ALEX REINHART
MARCH 2015, 176 PP., $24.95
ISBN 978-1-59327-620-1

AUTOMATE THE BORING STUFF WITH PYTHON
Practical Programming for Total Beginners
by AL SWEIGART
APRIL 2015, 504 PP., $29.95
ISBN 978-1-59327-599-0

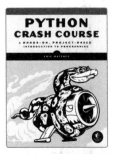

PYTHON CRASH COURSE
A Hands-On, Project-Based Introduction to Programming
by ERIC MATTHES
FALL 2015, 624 PP., $39.95
ISBN 978-1-59327-603-4

TEACH YOUR KIDS TO CODE
A Parent-Friendly Guide to Python Programming
by BRYSON PAYNE
APRIL 2015, 336 PP., $29.95
ISBN 978-1-59327-614-0

PYTHON PLAYGROUND
Geeky Weekend Projects for the Curious Programmer
by MAHESH VENKITACHALAM
FALL 2015, 304 PP., $29.95
ISBN 978-1-59327-604-1

PHONE:
800.420.7240 OR
415.863.9900

EMAIL:
SALES@NOSTARCH.COM

WEB:
WWW.NOSTARCH.COM